Electric Power Systems

A First Course

Electric Power Systems

A First Course

NED MOHAN
Oscar A. Schott Professor of Power Electronics and Systems
Department of Electrical and Computer Engineering
University of Minnesota

WILEY
John Wiley & Sons, Inc.

VP & PUBLISHER:	Don Fowley
EDITOR:	Dan Sayre
PROJECT EDITOR:	Nithyanand Rao
EDITORIAL ASSISTANT:	Charlotte Cerf
MARKETING MANAGER:	Christopher Ruel
MARKETING ASSISTANT:	Ashley Tomeck
DESIGNER:	James O'Shea
SENIOR PRODUCTION MANAGER:	Janis Soo
SENIOR PRODUCTION EDITOR:	Joyce Poh

PSpice is a registered trademark of the OrCAD Corporation.
SIMULINK is a registerd trademark of The Mathworks, Inc.

This book was set in 10/12 TimesNewRoman by MPS Limited, a Macmillan Company, Chennai, India and printed and bound by Hamilton Printing. The cover was printed by Hamilton Printing.

This book is printed on acid free paper.

Founded in 1807, John Wiley & Sons, Inc. has been a valued source of knowledge and understanding for more than 200 years, helping people around the world meet their needs and fulfill their aspirations. Our company is built on a foundation of principles that include responsibility to the communities we serve and where we live and work. In 2008, we launched a Corporate Citizenship Initiative, a global effort to address the environmental, social, economic, and ethical challenges we face in our business. Among the issues we are addressing are carbon impact, paper specifications and procurement, ethical conduct within our business and among our vendors, and community and charitable support. For more information, please visit our website: www.wiley.com/go/ citizenship.

Library of Congress Cataloging-in-Publication Data

Mohan, Ned.
    Electric power systems : a first course / Ned Mohan.
      p. cm.
    Includes bibliographical references and index.
    ISBN 978-1-118-07479-4 (acid free paper)
      1. Electric power systems. I. Title.
  TK1001.M5985   2012
  621.31—dc23

                         2011044321

Printed in the United States of America
10 9 8 7 6 5 4 3 2 1

CONTENTS

PREFACE

Role of Electric Power Systems in Sustainability:

It is estimated that in the United States, approximately forty percent of the energy used is first converted into electricity. This percentage would grow to sixty to seventy percent if we begin to use electricity for transportation by means of high-speed trains, and electric and electric-hybrid vehicles. Of course, generating electricity by using renewables and using it efficiently are both extremely important for sustainability. However, the electricity is often generated in areas remote to where it is used, and therefore how efficiently, and reliably, it is delivered is equally important for sustainability.

Lately there is a great deal of emphasis on smart grid whose definition remains somewhat vague. Nonetheless, we can all agree that it in a broad sense, it is to deliver electricity reliably and efficiently, to allow efficient end-use and to facilitate the integration of electricity harnessed from renewables such as solar and wind, and storage. To derive the benefits which a smart grid may potentially offer, a thorough understanding of how electric power networks operate is extremely important, and that is the purpose of this textbook.

The subject of Electric Power Systems encompasses a large and complex set of topics. The novel aspect of this textbook is a balanced approach in presenting as many topics as possible on a fundamental basis for a single-semester course. These topics include how electricity is generated, and how it is used by various loads, and the network and various apparatus in between. Students see the big picture and learn the fundamentals at the same time. Sequencing of these topics is considered carefully to avoid repetition and to retain student interest. However, instructors can rearrange the order for the most part, based on their own experiences and preferences.

In a fast-paced course such as this, student learning can be significantly enhanced by computer simulations. Therefore, simulation exercises and examples are included throughout this textbook using state-of-the-art simulation packages such as *MATLAB*, *Simulink*, *PowerWorld* and *PSCAD-EMTDC*.

SUGGESTED REFERENCE BOOKS:

Over the years, this author has immensely benefited from the following books (in no particular order) and their authors are my teachers. These are excellent books that are highly recommended as references in this course:

1. Electric Utility Systems and Practices, 4th Edition by Homer M. Rustebakke (Editor), John Wiley & Sons, August 1983.
2. Powerplant Technology by M. M. El-Wakil, McGraw-Hill Companies, 1984.

3. Electric Power Research Institute (EPRI), *Transmission Line Reference Book: 345 kV and above*, 2nd edition, 1982.

4. Hermann W. Dommel, *EMTP Theory Book*, BPA, August 1986.

5. Prabha Kundur, *Power System Stability and Control*, McGraw Hill, 1994.

6. Paul Anderson, Analysis of Faulted Power Systems, IEEE Press, 1995.

7. W. D. Stevenson, *Elements of Power System Analysis*, 4th edition, McGraw-Hill, 1982.

8. Electrical Transmission and Distribution Reference Book, Westinghouse Electric Corporation, 1950.

9. E. W. Kimbark, Direct Current Transmission, vol. 1, Wiley-Interscience, New York, 1971.

10. J. Casazza and F. Delea, Understanding Electric Power Systems: An Overview of the Technology and the Marketplace, IEEE Press and Wiley-Interscience, 2003.

11. C.W. Taylor, Power System Voltage Stability, McGraw-Hill, 1994 (for reprints, Email: cwtaylor@ieee.org).

12. N. Hingorani, L. Gyugyi, *Understanding FACTS : Concepts and Technology of Flexible AC Transmission Systems*, Wiley-IEEE Press, 1999.

13. Leon K. Kirchmayer, *Economic Operation of Power Systems*, John Wiley & Sons, 1958.

14. Nathan Cohn, *Control of Generation and Power Flow on Interconnected Systems*, John Wiley & Sons, 1967.

15. Leon K. Kirchmayer, *Economic Control of Interconnected Systems*, John Wiley & Sons, 1959.

16. A. Wood, B. Wollenberg, *Power Generation, Operation, and Control*, 2nd edition, Wiley-Interscience, 1996.

17. United States Department of Agriculture, Rural Utilities Service, Design Guide for Rural Substations, RUS BULLETIN 1724E-300 (http://www.rurdev.usda.gov/RDU_Bulletins_Electric.html).

1

POWER SYSTEMS: A CHANGING LANDSCAPE

Electric power systems are a technical wonder and, according to the National Academy of Engineering [1], electricity and its accessibility are the greatest engineering achievements of the twentieth century, ahead of computers and airplanes.

Electricity is a highly refined "commodity" without which it is difficult to imagine how a modern society could function. It has saved countless millions from the daily drudgery of backbreaking menial tasks. Electricity as a commodity is also unique in that it must be consumed at the instant it is produced, since storing it for future use is still economically prohibitive in most cases.

1.1 NATURE OF POWER SYSTEMS

Power systems encompass the generation of electricity to its ultimate consumption in operating of computers to hairdryers. For example, the North American grid in the United States and Canada consists of thousands of generators, all operating in synchronism. These generators are interconnected by over two hundred thousand miles of transmission lines at 230 kV and above [2], as shown in Figure 1.1. Such an interconnected system results in the continuity and the reliability of service if there is an outage in one part of the system, and provides electricity at the lowest cost by utilizing the lowest cost generation, as much as possible, at a given time.

Although power transmission systems are almost always three-phase, they are represented by the one-line diagram, as illustrated by Figure 1.2.

The output of generators, often in locations remote to the load centers, is stepped up in voltage from a 20-kV level by transformers to a much higher voltage for transmission over long distances, and subsequently stepped down in voltage level and distributed to the consumers. A visit to the local utility substation will show many of the apparatus in operation day and night. A similar visit to a generation plant and the control center of the local utility will illustrate how these interconnected systems are kept operating in the most economical and reliable manner.

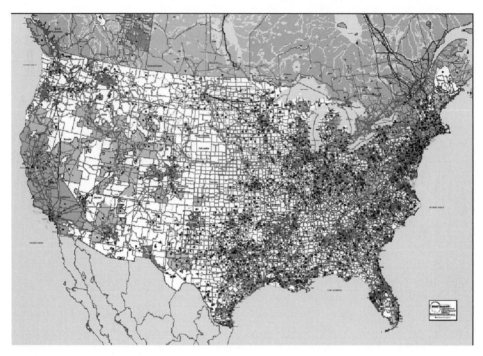

FIGURE 1.1   Interconnected North American power grid [2].

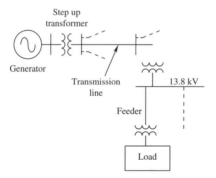

FIGURE 1.2   One-line diagram; the sub-transmission system is not shown.

1.2   CHANGING LANDSCAPE OF POWER SYSTEMS AND UTILITY DEREGULATION

Power systems of today are undergoing major changes in how they are evolving in their structure and how they are meeting the load demand. In the past and true to some extent today, electric utilities were highly centralized, owning large central power plants as well as the transmission and distribution systems, all the way down to the consumer loads. These utilities were monopolies where consumers had no choice but to buy power from their local utilities. For oversight purposes, utilities were highly regulated by public service commissions that acted as consumer watchdogs, preventing utilities from price gouging, and as custodians of the environment by not allowing avoidable polluting practices.

The structure and the operation of the power systems are beginning to change and the utilities have been divided into separate generation and transmission/distribution companies. There is distributed generation (DG) by independent power producers (IPP), who can generate electricity by whatever means, by wind for example, and who must be allowed access to the transmission grid to sell power to consumers. The impetus for the breakup in the utility structure was provided by the enormous benefits of deregulation in the telecommunication and the airline industries, which fostered a large degree of competition, resulting in much lower rates and much better service to the consumers. In spite of the inherent differences between these two industries and the utility industry, it was perceived that the utility deregulation would similarly profit the consumers with lower electricity rates. This deregulation is in transition, with some states and countries pursuing it more aggressively, and others more cautiously. To promote open competition, utilities are forced to restructure by unbundling their generation units from their transmission and distribution units. The objective is that the independent transmission system operators (TSO) wheel power for a charge from anywhere and from anyone to the customer site. This would foster competition, allowing open transmission access to everyone, for example, independent power producers (IPP). Many such small IPPs have gone into business, producing power using gas turbines and windmills.

Operation in a reliable manner is ensured by the Independent System Operators (ISOs) and the financial transactions are governed by real-time bidding to buy and sell power. Energy traders have gotten into the act for profit: buying energy at lower prices and selling it at higher prices in the spot market. Utilities are signing long-term contracts for energy such as gas. This is all based on the rules of the financial world: forecasting, risks, options, reliability, and so forth.

As mentioned earlier, the outcome of this deregulation, still in transition, is far from certain. So far, consumers have seen little or no savings and in some cases it has been a disaster as during the California energy crisis. However, there is every reason to believe that the deregulation, now in progress, would continue with very little possibility that the clock would be turned back. In doing so, there are some fixes that are needed. The transmission grid has become a bottleneck with very little financial incentive for the transmission system operators to increase the capacity. If the transmission system is congested, TSOs can charge higher prices. The number of transactions and the complexity of these transactions have increased drastically. These factors point to anticipated legislative actions needed to maintain electric system reliability.

1.3 TOPICS IN POWER SYSTEMS

The purpose of this textbook is to provide a complete overview of power systems meeting present and future energy challenges. In doing so, fundamentals are to be clearly understood so that the details in any of the topics can be easily comprehended as a graduate student or as a lifelong learner while practicing this profession.

The chapters in this book are arranged to associate the lecture material chronologically with the laboratory exercises.

As the present chapter describes, the power systems industry is undergoing major changes. In order to fully comprehend these changes and the underlying challenges, we need to understand the basics of power systems. The most basic aspect of power systems is generation. One has a choice of choosing certain resources for generating electricity, but there are always consequences to any selection. These are briefly discussed in Chapter 3, which also describes the increasing role of renewable energy sources such as wind. However, prior to this, basic concepts that are fundamental to the analysis of power system circuits are briefly reviewed in Chapter 2.

AC transmission lines and cables are described in Chapter 4. Increasingly, high-voltage DC transmission (HVDC) is being used to enhance system stability. For the purposes of planning and operating securely under contingencies caused by outages, it is important to know how power flows on various paths. The principles of power flow are examined in Chapter 5. Voltages produced by generators are stepped up by transformers for long distance hauling of power over transmission lines. Transformers are described in Chapter 6. The principles behind HVDC are described in Chapter 7. Chapter 8 describes consumer loads and the role of power electronics that is changing their nature. This chapter also describes how these loads react to voltage fluctuations and the power quality.

For generating electricity, steam and natural gas are utilized to run turbines that provide mechanical input to synchronous generators to produce three-phase electrical voltages. Synchronous generators are described in Chapter 9. Transmission lines are being loaded more than ever, making the voltage stability a concern, as discussed in Chapter 10. There is a growing role of power electronics in power systems in the form of flexible AC transmission systems (FACTS), which are described in this chapter to improve the voltage stability.

Maintaining stability so that various generators operate in synchronism is described in Chapter 11, which discusses how the stability in an interconnected system, with thousands of generators operating in synch, can be maintained in response to transient conditions, such as transmission-line faults, where there is a mismatch between the mechanical power input to the turbines and the electrical power that can be transmitted. Chapter 12 discusses economic dispatch where generators are loaded such as to provide the overall economy of operation. The operation of interconnected systems, so that the power system frequency and voltages are maintained at their nominal values and the purchasing-and-selling agreements between various utilities are honored, are also described in Chapter 12.

Power systems are spread over large areas. Being exposed to the elements of nature, they are subjected to occasional faults against which they must be protected by design so that such events result only in momentary loss of power, and no permanent equipment damage accrues. Short-circuits on transmission systems is discussed in Chapter 13, which describes how relays detect faults and cause circuit breakers to open the circuit, inter-rupting the fault current, and then reclose to bring the operation back to normal as soon as possible. Lightening strikes and switching of extra-high-voltage transmission lines during reenergizing, particularly with trapped charge, can result in very-high-voltage surges which can cause insulation to flash over. To avoid this, surge arresters are used and are properly coordinated with the insulation level of the power systems apparatus to prevent damage. These topics are discussed in Chapter 14.

REFERENCES

1. National Academy of Engineering (www.nae.edu).
2. PennWell MAPSearch (www.mapsearch.com/paper_products.cfm).

PROBLEMS

1.1   What are the advantages of a highly interconnected system?
1.2   What changes are taking place in the utility industry?
1.3   What are the different topics in power systems necessary for understanding its basic nature?

2

REVIEW OF BASIC ELECTRIC CIRCUITS AND ELECTROMAGNETIC CONCEPTS

2.1 INTRODUCTION [1]

The purpose of this chapter is to review elements of basic electric circuit theory that are essential to the study of electric power circuits: use of phasors to analyze circuits in sinusoidal steady state, the real and reactive powers, the power factor, analysis of three-phase circuits, power flow in AC circuits and per-unit quantities.

In this book, we will use MKS units and the IEEE-standard letters and graphic symbols whenever possible. The lowercase letters v and i are used to represent instantaneous values of voltages and currents that vary as functions of time. A current's positive direction is indicated by an arrow, as shown in Figure 2.1. Similarly, the voltage polarities must be indicated. The voltage v_{ab} refers to the voltage of node a with respect to node b; thus $v_{ab} = v_a - v_b$.

2.2 PHASOR REPRESENTATION IN SINUSOIDAL STEADY STATE

In linear circuits with sinusoidal voltages and currents of frequency f applied for a long time so that steady state has been reached, all circuit voltages and currents are at a

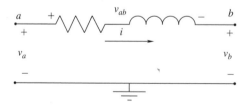

FIGURE 2.1 Convention for voltages and currents.

6

frequency $f(=\omega/2\pi)$. To analyze such circuits, calculations are simplified by means of phasor-domain analysis. The use of phasors also provides a deeper insight with relative ease into the circuit behavior.

In the phasor domain, the time-domain variables $v(t)$ and $i(t)$ are transformed into phasors which are represented by the complex variables $\overline{V}$ and $\overline{I}$. Note that these phasors are expressed by uppercase letters with a bar "−" on top. In a complex (real and imaginary) plane, these phasors can be drawn with a magnitude and an angle. A co-sinusoidal time function is taken as a reference phasor which is entirely real with an angle of zero degrees. Therefore, the time-domain voltage expression in Equation 2.1 below is represented by a corresponding phasor

$$v(t) = \sqrt{2}\,V\cos(\omega t + \phi_v) \quad \Leftrightarrow \quad \overline{V} = V \angle \phi_v \tag{2.1}$$

Similarly,

$$i(t) = \sqrt{2}\,I\cos(\omega t + \phi_i) \quad \Leftrightarrow \quad \overline{I} = I \angle \phi_i \tag{2.2}$$

where V and I are the RMS values of the voltage and current. These voltage and current phasors are drawn in Figure 2.2. In Equations 2.1 and 2.2, the angular frequency ω is implicitly associated with each phasor. Knowing this frequency, a phasor expression can be re-transformed back into a time-domain expression.

Using phasors, we can convert differential equations into easily solvable algebraic equations containing complex variables. Consider the circuit of Figure 2.3a in a sinusoidal steady state with an applied voltage at a frequency $f(=\omega/2\pi)$.

In order to calculate the current in this circuit, remaining in time domain would require us to solve the following differential equation:

$$Ri(t) + L\frac{di(t)}{dt} + \frac{1}{C}\int i(t)\cdot dt = \sqrt{2}\,V\cos(\omega t) \tag{2.3}$$

Using phasors, we can redraw the circuit of Figure 2.3a in Figure 2.3b, where the inductance L is represented by $j\omega L$ and the capacitance C by $j\left(\frac{1}{-\omega C}\right)$.

In the phasor-domain circuit, the impedance Z of the series-connected elements is obtained by the impedance triangle of Figure 2.3c as

$$Z = R + jX_L + jX_c \tag{2.4}$$

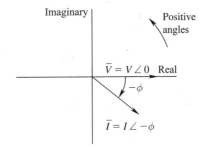

FIGURE 2.2 Phasor diagram.

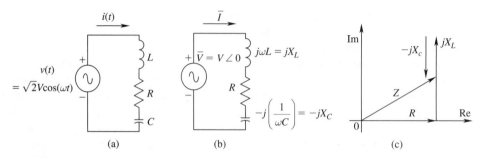

FIGURE 2.3   A circuit (a) in time-domain and (b) in phasor-domain; (c) impedance triangle.

where the reactance X is the imaginary part of an impedance Z:

$$X_L = \omega L, \quad \text{and} \quad X_c = \left(\frac{1}{-\omega C} \right) \tag{2.5}$$

This impedance can be expressed as

$$Z = |Z| \angle \phi \tag{2.6a}$$

where

$$|Z| = \sqrt{R^2 + \left(\omega L - \frac{1}{\omega C} \right)^2} \quad \text{and} \quad \phi = \tan^{-1} \left[\frac{\left(\omega L - \frac{1}{\omega C} \right)}{R} \right] \tag{2.6b}$$

It is important to recognize that while Z is a complex quantity, it is *not* a phasor and does *not* have a corresponding time-domain expression.

Example 2.1
Calculate the impedance seen from the terminals of the circuit in Figure 2.4 under a sinusoidal steady state at a frequency $f = 60$ Hz.

Solution $Z = j0.1 + \frac{2 \times (-j5)}{(2 - j5)} = 1.72 - j0.59 = 1.82 \angle -18.9° \, \Omega.$

Using the impedance in Equation 2.6, and assuming that the voltage phase angle ϕ_v is zero, the current in Figure 2.3b can be obtained as

$$\bar{I} = \frac{\bar{V}}{Z} = \left(\frac{V}{|Z|} \right) \angle -\phi \tag{2.7}$$

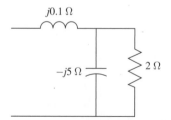

FIGURE 2.4   Impedance network of Example 2.1.

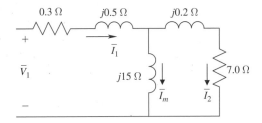

FIGURE 2.5 Circuit of Example 2.2.

where $I = \frac{V}{|Z|}$ and ϕ is as calculated from Equation 2.6b. Using Equation 2.2, the current can be expressed in the time domain as

$$i(t) = \frac{\sqrt{2}V}{|Z|}\cos(\omega t - \phi) \tag{2.8}$$

In the impedance triangle of Figure 2.3c, a positive value of the phase angle ϕ implies that the current lags behind the voltage in the circuit of Figure 2.3a. Sometimes, it is convenient to express the inverse of the impedance, which is called admittance:

$$Y = \frac{1}{Z} \tag{2.9}$$

The phasor-domain procedure for solving $i(t)$ is much easier than solving the differential-integral equation given by Equation 2.3.

Example 2.2

Calculate the current $\overline{I}_1$ and $i_1(t)$ in the circuit of Figure 2.5 if the applied voltage has an rms value of 120 V and a frequency of 60 Hz. Assume $\overline{V}_1$ to be the reference phasor.

Solution   With $\overline{V}_1$ as the reference phasor, it can be written as $\overline{V}_1 = 120 \angle 0° \, V$. The input impedance of the circuit seen from the applied voltage terminals is

$$Z_{in} = (0.3 + j0.5) + \frac{(j15)(7 + j0.2)}{(j15) + (7 + j0.2)} = 6.775 \angle 29.03° \, \Omega.$$

Therefore, the current $\overline{I}_1$ can be obtained as

$$\overline{I}_1 = \frac{\overline{V}_1}{Z_{in}} = \frac{120 \angle 0°}{6.775 \angle 29.03°} = 17.712 \angle -29.03° \, A$$

hence $i_1(t) = 25.048 \cos(\omega t - 29.03°) \, A.$

2.3   POWER, REACTIVE POWER, AND POWER FACTOR

Consider the generic circuit of Figure 2.6 in a sinusoidal steady state. Each subcircuit may consist of passive (R-L-C) elements and active voltage and current sources. Based on the arbitrarily chosen voltage polarity and the current direction shown in Figure 2.6,

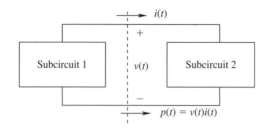

FIGURE 2.6   A generic circuit divided into two subcircuits.

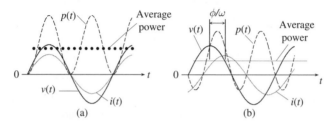

FIGURE 2.7   Instantaneous power with sinusoidal currents and voltages.

the instantaneous power $p(t) = v(t)i(t)$ is delivered by subcircuit 1 and absorbed by subcircuit 2. This is because in subcircuit 1, the positively defined current is coming out of the positive-polarity terminal (the same as in a generator). On the other hand, the positively defined current is entering the positive-polarity terminal in subcircuit 2 (the same as in a load). A negative value of $p(t)$ reverses the roles of subcircuit 1 and subcircuit 2.

Under a sinusoidal steady state condition at a frequency f, the complex power S, the reactive power Q, and the power factor express how effectively the real (average) power P is transferred from one sub-circuit to the other. If $v(t)$ and $i(t)$ are in phase, the instantaneous power $p(t) = v(t)i(t)$, as shown in Figure 2.7a, pulsates at twice the steady state frequency as shown below (V and I are the RMS values):

$$p(t) = \sqrt{2}V\cos\omega t \cdot \sqrt{2}I\cos\omega t = 2\,VI\cos^2\omega t = VI + VI\cos 2\omega t \quad (i \text{ in phase with } v) \quad (2.10)$$

where both ϕ_v and ϕ_i are assumed to be zero without any loss of generality. In this case, at all times, $p(t) \geq 0$, and therefore the power always flows in one direction: from subcircuit 1 to subcircuit 2. The average over one cycle of the second term on the right side of Equation 2.10 is zero, and therefore the average power is $P = VI$.

Now consider the waveforms of Figure 2.7b, where the $i(t)$ waveform lags behind the $v(t)$ waveform by a phase angle $\phi(t)$. Here, $p(t)$ becomes negative during a time interval of (ϕ/ω) during each half-cycle as calculated below:

$$p(t) = \sqrt{2}V\cos\omega t \cdot \sqrt{2}I\cos(\omega t - \phi) = VI\cos\phi + VI\cos(2\omega t - \phi) \quad (2.11)$$

A negative instantaneous power implies power flow in the opposite direction. This back-and-forth flow of power indicates that the real (average) power is not optimally trans-ferred from one subcircuit to the other, as is the case in Figure 2.7a. Therefore, the average power $P(= VI\cos\phi)$ in Figure 2.7b is less than that in Figure 2.7a.

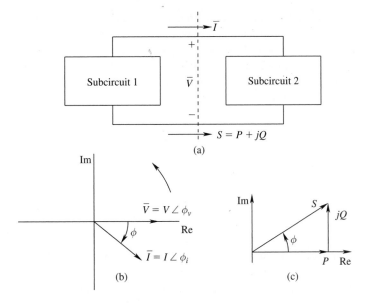

FIGURE 2.8 (a) Circuit in phasor-domain; (b) phasor diagram with $\phi_v = 0$; (c) power triangle.

The circuit of Figure 2.6 is redrawn in Figure 2.8a in the phasor domain. The voltage and the current phasors are defined by their magnitudes and phase angles as

$$\overline{V} = V \angle \phi_v \quad \text{and} \quad \overline{I} = I \angle \phi_i \tag{2.12}$$

The complex power S is defined as

$$S = \overline{V}\,\overline{I}^* \quad (\text{* indicates complex conjugate}) \tag{2.13}$$

Therefore, substituting the expressions for voltage and current into Equation 2.13, and noting that $\overline{I}^* = I \angle -\phi_i$,

$$S = V \angle \phi_v \; I \angle -\phi_i = VI \angle (\phi_v - \phi_i) \tag{2.14}$$

The difference between the two phase angles is defined as

$$\phi = \phi_v - \phi_i \tag{2.15}$$

Therefore,

$$S = VI \angle \phi = P + jQ \tag{2.16}$$

where

$$P = VI \cos \phi \tag{2.17}$$

$$Q = VI \sin \phi \tag{2.18}$$

In Equation 2.17, $I\cos\phi$ is the current component that is in phase with the voltage phasor in Figure 2.8b, and results in real power transfer P. Whereas, from Equation 2.18, $I\sin\phi$ is the current component that is at 90° angle to the voltage phasor in Figure 2.8b, and results in a reactive power Q but no average real power.

The power triangle corresponding to the phasors in Figure 2.8b is shown in Figure 2.8c. From Equation 2.16, the magnitude of S, also called the apparent power, is

$$|S| = \sqrt{P^2 + Q^2} \tag{2.19}$$

and

$$\phi = \tan^{-1}\left(\frac{Q}{P}\right) \tag{2.20}$$

The above quantities have the following units: P: W (Watts); Q: VAR (Volt-Amperes Reactive); $|S|$: VA (Volt-Amperes); and ϕ_v, ϕ_i, ϕ: radians, measured positively in a counterclockwise direction with respect to the real-axis that is drawn horizontally from left to right.

The physical significance of the apparent power $|S|$, P, and Q should be understood. The cost of most electrical equipment such as generators, transformers, and transmission lines is proportional to $|S|$ $(= VI)$, since their electrical insulation level and the magnetic core size for a given line-frequency depend on the voltage V and the conductor size depends on the rms current I. The real power P has a physical significance since it represents the useful work being performed plus the losses. Under most operating conditions, it is desirable to have the reactive power Q be zero, or it results in increased $|S|$.

To support the above discussion, another quantity called the power factor is defined. The power factor is a measure of how effectively a load draws real power:

$$\text{power factor} = \frac{P}{|S|} = \frac{P}{VI} = \cos\phi \tag{2.21}$$

which is a dimensionless quantity. Ideally, the power factor should be 1.0 (that is, Q should be zero) in order to draw real power with a minimum current magnitude and hence minimize losses in electrical equipment, and transmission and distribution lines. An inductive load draws power at a lagging power factor where the load current lags behind the voltage. Conversely, a capacitive load draws power at a leading power factor where the load current leads the load voltage.

Example 2.3
Calculate P, Q, S, and the power factor of operation at the terminals in the circuit of Figure 2.5 in Example 2.2.

Solution

$$P = V_1 I_1 \cos\phi = 120 \times 17.712 \cos 29.03° = 1858.4\,\text{W}$$
$$Q = V_1 I_1 \sin\phi = 120 \times 17.712 \times \sin 29.03° = 1031.3\,\text{VAR}$$
$$|S| = V_1 I_1 = 120 \times 17.7 = 2125.4\,\text{VA}$$

From Equation 2.20, $\phi = \tan^{-1}\frac{Q}{P} = 29.03°$ in the power triangle of Figure 2.8c. Note that the angle of S in the power triangle is the same as the impedance angle ϕ in Example 2.2. The power factor of operation is

$$\text{power factor} = \cos\phi = 0.874$$

Note the following for the inductive impedance in the above example: (1) The impedance is $Z = |Z|\angle\phi$, where ϕ is positive. (2) The current lags the voltage by the impedance angle ϕ. This corresponds to a lagging power factor of operation. (3) In the power triangle, the impedance angle ϕ relates P, Q, and S. (4) An inductive impedance, when applied a voltage, draws a positive reactive power Q_L (vars). If the impedance were to be capacitive, the phase angle ϕ would be negative and this impedance, when applied a voltage, would draw a negative reactive power Q_C (that is, this impedance would *supply* a positive reactive power).

2.3.1 Sum of Real and Reactive Powers in a Circuit

In a circuit, all the real powers supplied to the various components sum to the total real power supplied:

$$\text{Total Real Power Supplied} = \sum_k P_k = \sum_k I_k^2 R_k \tag{2.22}$$

Similarly, all the reactive powers supplied to the various components sum to the total reactive power supplied:

$$\text{Total Reactive Power Supplied} = \sum_k Q_k = \sum_k I_k^2 X_k \tag{2.23}$$

where X_k and hence Q_k is negative if a component capacitive.

Example 2.4

In the circuit of Figure 2.5 in Example 2.2, calculate P and Q associated with each element and calculate the total real and reactive powers supplied at the terminals. Confirm these calculations by comparing with P and Q calculated in Example 2.3.

Solution   From Example 2.2, $\bar{I}_1 = 17.712\angle -29.03°$ A.

$$\bar{I}_m = \bar{I}_1 \frac{R_2 + jX_2}{(R_2 + jX_2) + jX_m} = 7.412\angle -92.66° \text{ A}, \text{ and } \bar{I}_2 = \bar{I}_1 - \bar{I}_m = 15.876\angle -4.3° \text{ A}.$$

Therefore, $P_{R_1} = R_1 I_1^2 = 0.3 \times 17.172^2 = 94.11$ W, $P_{R_2} = R_2 I_2^2 = 7 \times 15.876^2 = 1764.3$ W, and

$$\sum P = P_{R_1} + P_{R_2} = 1858.4 \text{ W}$$

For the reactive vars, $Q_{X_1} = X_1 I_1^2 = 0.5 \times 17.172^2 = 156.851$ VAR,

$$Q_{X_2} = X_2 I_2^2 = 0.2 \times 15.876^2 = 50.409 \text{ VAR}, \quad \text{and}$$

$$Q_{X_3} = X_m I_m^2 = 15 \times 7.412^2 = 821.021 \text{ VAR}$$

Therefore,

$$\sum Q = Q_{X_1} + Q_{X_2} + Q_{X_m} = 1031.3 \text{ VAR}$$

Note that $\sum P$ and $\sum Q$ are equal to P and Q at the terminals, as calculated in Example 2.3.

2.3.2 Power Factor Correction

As explained earlier, utilities prefer that loads draw power at the unity power factor, so that the current for a given power drawn is minimum, thus resulting in the least amount of $I^2 R$ losses in the resistances associated with transmission and distribution lines and other apparatus. This power factor correction can be accomplished by compensating or nullifying the reactive power drawn by the load by connecting a reactance in parallel that draws the same reactive power in magnitude but of opposite sign. This is illustrated by the example below.

Example 2.5
In the circuit of Figure 2.5 and Example 2.3, the complex power drawn by the load impedance was calculated as $P_L + jQ_L = (1858.4 + j1031.3)$ VA. Calculate the capacitive reactance in parallel that will result in the overall power factor to be unity, as seen from the voltage source.

Solution   The load is drawing reactive power $Q_L = 1031.3$ VAR. Therefore, as shown in Figure 2.9, a capacitive reactance must be connected in parallel with the load impedance to draw $Q_C = -1031.3$ VAR (or *supply* positive reactive vars equal to Q_L) such that only the real power is drawn from the voltage source.

Since the voltage across the capacitive reactance is given, the capacitive reactance can be calculated as

$$X_C = \frac{V^2}{Q_C} = \frac{120^2}{-1031.3} = -13.96 \ \Omega.$$

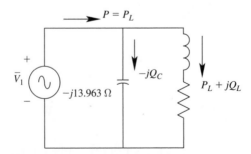

FIGURE 2.9   Power factor correction in Example 2.5.

2.4 THREE-PHASE CIRCUITS

Basic understanding of three-phase circuits is essential to the study of electric power systems. Nearly all electricity, with only a few exceptions, is generated by means of three-phase AC generators. Figure 2.10 shows a simplified one-line diagram of a three-phase power system. Generated voltages (usually below 25 kV) are stepped-up by means of transformers to 230 kV to 500 kV levels for transferring power long distances over transmission lines from the generation site to load centers. At the receiving end of the transmission lines near load centers, these three-phase voltages are stepped-down by transformers. Most motor loads above a few kW in power rating operate from three-phase voltages.

Three-phase AC circuits are either wye or delta connected. We will investigate these under sinusoidal steady state, balanced condition, which implies that all three voltages are equal in magnitude and displaced by 120° ($2\pi/3$ radians) with respect to each other. The phase sequence of voltages is commonly assumed to be $a - b - c$, where the phase a voltage leads the phase b voltage by 120°, and phase b leads phase c by 120° ($2\pi/3$ radians), as shown in Figure 2.11. This applies to both the time-domain representation in Figure 2.11a and the phasor-domain representation in Figure 2.11b. Notice that in the $a - b - c$ sequence voltages plotted in Figure 2.11a, first v_{an} reaches its positive peak, and then v_{bn} reaches its positive peak $2\pi/3$ radians later, and so on. We can represent these voltages in the phasor form in Figure 2.11b as

$$\overline{V}_{an} = V_{ph} \angle 0°, \qquad \overline{V}_{bn} = V_{ph} \angle -120°, \quad \text{and} \qquad \overline{V}_{cn} = V_{ph} \angle -240° \qquad (2.24)$$

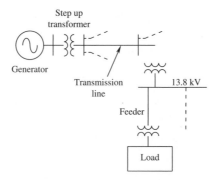

FIGURE 2.10 One-line diagram of a three-phase power system; the sub-transmission is not shown.

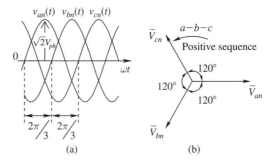

FIGURE 2.11 Three-phase voltages in time and phasor domain.

where V_{ph} is the RMS phase-voltage magnitude and the phase a voltage is assumed to be the reference (with an angle of zero degrees).

For a balanced set of voltages given by Equation 2.24, at any instant, the sum of these phase voltages equals zero:

$$\overline{V}_{an} + \overline{V}_{bn} + \overline{V}_{cn} = 0 \quad \text{and} \quad v_{an}(t) + v_{bn}(t) + v_{cn}(t) = 0 \tag{2.25}$$

2.4.1   Per-Phase Analysis in Balanced Three-Phase Circuits

A three-phase circuit can be analyzed on a per-phase basis, provided that it has a balanced set of source voltages and equal impedances in each of the phases. Such a wye-connected circuit is shown in Figure 2.12a.

In such a circuit, the source neutral n and the load neutral N are at the same potential. Therefore, hypothetically connecting these with a zero impedance wire, as shown in Figure 2.12b, does not change the original three-phase circuit, which can now be analyzed on a per-phase basis. Selecting phase a for this analysis, the per-phase circuit is shown in Figure 2.13a.

If $Z_L = |Z_L| \angle \phi$, using the fact that in a balanced three-phase circuit, phase quantities are displaced by 120° with respect to each other, we find that

$$\overline{I}_a = \frac{\overline{V}_{an}}{Z_L} = \frac{V_{ph}}{|Z_L|} \angle - \phi \quad \overline{I}_b = \frac{\overline{V}_{bn}}{Z_L} = \frac{V_{ph}}{|Z_L|} \angle \left(-\phi - \frac{2\pi}{3} \right) \quad \overline{I}_c = \frac{\overline{V}_{cn}}{Z_L}$$

$$= \frac{V_{ph}}{|Z_L|} \angle \left(-\phi - \frac{4\pi}{3} \right) \tag{2.26}$$

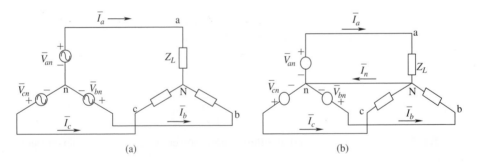

FIGURE 2.12   Balanced wye-connected, three-phase circuit.

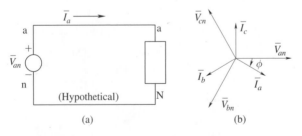

FIGURE 2.13   Per-phase circuit and the corresponding phasor diagram.

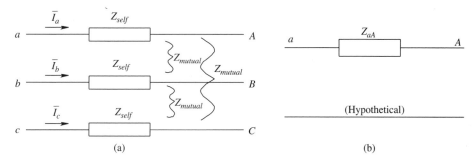

FIGURE 2.14   Balanced three-phase network with mutual couplings.

The three-phase voltage and current phasors are shown in Figure 2.13b. The total real and reactive powers in a balanced three-phase circuit can be obtained by multiplying the per-phase values by a factor of three. The total power factor is the same as its per-phase value.

2.4.2   Per-Phase Analysis of Balanced Circuits including Mutual Couplings

Most three-phase apparatus such as generators, transmission lines and motors consist of mutually coupled phases. For example, in a three-phase synchronous generator, the current in a phase winding produces flux lines that link not only that phase winding but the other phase windings as well. In general, in a balanced three-phase circuit, this can be represented as shown in Figure 2.14a where Z_{self} is the impedance of a phase by itself, and Z_{mutual} represents mutual coupling. Therefore,

$$\overline{V}_{aA} = Z_{self}\overline{I}_a + Z_{mutual}\overline{I}_b + Z_{mutual}\overline{I}_c \tag{2.27}$$

In a balanced three-phase circuit under balanced excitation, three currents sum to zero: $\overline{I}_a + \overline{I}_b + \overline{I}_c = 0$. Therefore, making use of this condition in Equation 2.25,

$$\overline{V}_{aA} = (Z_{self} - Z_{mutual})\overline{I}_a = Z_{aA}\overline{I}_a \tag{2.28}$$

where,

$$Z_{aA} = Z_{self} - Z_{mutual} \tag{2.29}$$

and the per-phase representation is as shown in Figure 2.14b.

2.4.3   Line-to-Line Voltages

In the balanced wye-connected circuit of Figure 2.12a, it is often necessary to consider the line-to-line voltages, such as those between phases a and b, and so on. Based on the previous analysis, we can refer to both neutral points n and N by a common term n, since the potential difference between n and N is zero. Therefore, in Figure 2.12a, as shown in the phasor diagram of Figure 2.15,

$$\overline{V}_{ab} = \overline{V}_{an} - \overline{V}_{bn}, \qquad \overline{V}_{bc} = \overline{V}_{bn} - \overline{V}_{cn}, \quad \text{and} \quad \overline{V}_{ca} = \overline{V}_{cn} - \overline{V}_{an} \tag{2.30}$$

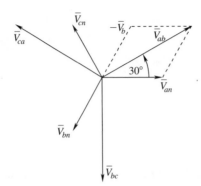

FIGURE 2.15   Line-to-line voltages in a three-phase circuit.

Either using Equations 2.24 and 2.30, or graphically from Figure 2.15, we can show the following where V_{ph} is the RMS magnitude of each of the phase voltages:

$$\overline{V}_{ab} = \sqrt{3}V_{ph} \angle \frac{\pi}{6}$$

$$\overline{V}_{bc} = \sqrt{3}V_{ph} \angle \left(\frac{\pi}{6} - \frac{2\pi}{3} \right) = \sqrt{3}V_{ph} \angle -\frac{\pi}{2} \tag{2.31}$$

$$\overline{V}_{ca} = \sqrt{3}V_{ph} \angle \left(\frac{\pi}{6} - \frac{4\pi}{3} \right) = \sqrt{3}V_{ph} \angle -\frac{7\pi}{6}$$

Comparing Equations 2.24 and 2.31, we see that the line-to-line voltages have an RMS value that is $\sqrt{3}$ times the phase voltage (RMS):

$$V_{LL} = \sqrt{3}V_{ph} \tag{2.32}$$

and $\overline{V}_{ab}$ leads $\overline{V}_{an}$ by $\pi/6$ radians (30°).

Example 2.6
In residential buildings where three-phase voltages are brought in, the RMS value of the line-line voltage is $V_{LL} = 208$ V. Calculate the RMS value of the phase voltages.

Solution   From Equation 2.32,

$$V_{ph} = \frac{V_{LL}}{\sqrt{3}} = 120 \text{ V}$$

2.4.4   Delta-Connection in AC Machines and Transformers

So far it is assumed that the three-phase sources and loads are connected in a wye configuration, as shown in Figure 2.12a. However, in AC machines and transformers, the three phase-windings may be connected in a delta configuration. The relationship between the line-currents and phase-currents under a balanced condition is described in Appendix 2A.1.

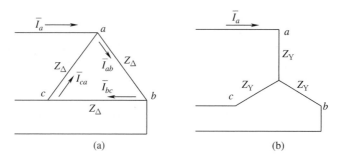

FIGURE 2.16 Delta-wye transformation of impedances under the balanced condition.

2.4.4.1 Delta-Wye Transformation of Load Impedances under Balanced Conditions

It is possible to replace the delta-connected load impedances by the equivalent wye-connected load impedances, and vice versa. Under a totally balanced condition, the delta-connected load impedances of Figure 2.16a can be replaced by the equivalent wye-connected load impedances of Figure 2.16b, similar to that in Figure 2.12a. We can then apply a per-phase analysis using Figure 2.13.

Under the balanced condition where all three impedances are the same, as shown in Figure 2.16, the impedance between terminals a and b, with the terminal c not connected, is $(2/3)Z_\Delta$ in the delta-connected circuit and $2Z_Y$ in the wye-connected circuit. Equating these two impedances,

$$Z_Y = \frac{Z_\Delta}{3} \tag{2.33}$$

The delta-wye transformation of unbalanced load impedances is analyzed in Appendix 2A.2.

2.4.5 Power, Reactive Power, and Power Factor in Three-Phase Circuits

We saw earlier that the per-phase analysis is valid for balanced three-phase circuits in sinusoidal steady state. This implies that the real and reactive powers drawn by each phase are the same as if were a single-phase load. Therefore, the total average real power and the reactive power in a three-phase circuit are ($V = V_{ph}$ and $I = I_{ph}$)

$$P_{3-phase} = 3 \times P_{per-phase} = 3VI \cos \phi \tag{2.34}$$

$$Q_{3-phase} = 3 \times Q_{per-phase} = 3VI \sin \phi \tag{2.35}$$

and the total apparent volt-ampers are

$$|S|_{3-phase} = 3 \times |S_{per-phase}| = 3VI \tag{2.36}$$

Therefore, the power factor in a three-phase circuit is the same as the per-phase power factor

$$\text{power factor} = \frac{P_{3-phase}}{|S|_{3-phase}} = \frac{3VI \cos \phi}{3VI} = \cos \phi \tag{2.37}$$

However, there is one very important difference between three-phase and single-phase circuits in terms of the instantaneous power. In both circuits, in each phase, the instantaneous power is pulsating as given by Equation 2.11 and repeated below for phase-a, where the phase current lags by a phase angle $\phi(t)$ behind the phase voltage, which is considered to be the reference phasor:

$$p_a(t) = \sqrt{2}V \cos \omega t \cdot \sqrt{2}I \cos(\omega t - \phi) = VI \cos\phi + VI \cos(2\omega t - \phi) \qquad (2.38)$$

Similar expressions can be written for phases b and c:

$$p_b(t) = \sqrt{2}V \cos(\omega t - 2\pi/3) \cdot \sqrt{2}I \cos(\omega t - \phi - 2\pi/3)$$
$$= VI \cos\phi + VI \cos(2\omega t - \phi - 4\pi/3) \qquad (2.39)$$

$$p_c(t) = \sqrt{2}V \cos(\omega t - 4\pi/3) \cdot \sqrt{2}I \cos(\omega t - \phi - 4\pi/3)$$
$$= VI \cos\phi + VI \cos(2\omega t - \phi - 8\pi/3) \qquad (2.40)$$

Addition of the three instantaneous powers in Equations 2.38 through 2.40 results in

$$p_{3-phase}(t) = 3VI \cos \phi = P_{3-phase} \quad \text{(from Equation 2.34)} \qquad (2.41)$$

which shows that the combined three-phase power in steady state is a constant, equal to its average value, even on an instantaneous basis. This is in contrast to the pulsating power, shown in Figure 2.7, in single-phase circuits. The non-pulsating total instantaneous power in three-phase circuits results in non-pulsating torque in motors and generators, and is the reason for preferring three-phase motors and generators over their single-phase counterparts.

Example 2.7
In the three-phase circuit of Figure 2.12a, $V_{LL} = 208$ V, $|Z_L| = 10\ \Omega$, and the per-phase power factor is 0.8 (lagging). Calculate the capacitive reactance needed, in parallel with the load impedance in each phase, to make the power factor 0.95 (lagging).

Solution The three-phase circuit of Figure 2.12a can be represented by the per-phase circuit of Figure 2.13a. Assuming the input voltage to the reference phasor, $\overline{V}_{an} = \frac{208}{\sqrt{3}} \angle 0 = \underbrace{120}_{(=V)} \angle 0$ V. The current $\overline{I}_L$ drawn by the load lags the voltage by an angle $\phi_L = \cos^{-1}(0.8) = 36.87°$, and

$$\overline{I}_L = \frac{V(=120)}{|Z_L|} \angle -\phi_L = 12 \angle -36.87° \text{ A.}$$

Therefore, the per-phase real power P_L, and the reactive var (volt-amperes reactive) Q_L drawn by the load are

$$P_L = \frac{V^2}{|Z_L|} \text{(power factor)} = 1152 \text{ W} \quad \text{and} \quad Q_L = \frac{V^2}{|Z_L|} \sin \phi_L = 864 \text{ VAR.}$$

To make the net power factor is to be 0.95 (lagging), a power-factor-correction capacitor of an appropriate value is connected in parallel with the load in each phase. Now, the net current into the combination of the load impedance and the power-factor-correction capacitor is at an angle of $\phi_{net} = \cos^{-1}(0.95) = 18.195°$(lagging). The net real power P_{net} drawn from the source still equals P_L, that is, $P_{net} = P_L$. Using the power-triangle of Figure 2.8c, the net reactive var drawn from the source is

$$Q_{net} = P_{net}\tan(\phi_{net}) = 378.65 \text{ VAR}.$$

Since $Q_{net} = Q_L - Q_{cap}$,

$$Q_{cap} = Q_L - Q_{net} = 864.0 - 378.65 = 485.35 \text{ VAR}.$$

Therefore, the capacitive reactance needed in parallel is

$$X_{cap} = \frac{V^2}{Q_{cap}} = 29.67 \ \Omega.$$

2.5 REAL AND REACTIVE POWER TRANSFER BETWEEN AC SYSTEMS

In this course, it will be important to calculate power flow between AC systems connected by transmission lines. Simplified AC systems can be represented by two AC voltage sources of the same frequency connected through a reactance X in series as shown in Figure 2.17a, where the series resistance has been neglected for simplification.

The phasor diagram for the system of Figure 2.17a, where the voltage $\overline{V}_R$ is assumed to be the reference voltage with its phase angle to be zero, is shown in Figure 2.17b. Based on the load, the current may be at some arbitrary phase angle ϕ. In the circuit of Figure 2.17a,

$$\overline{I} = \frac{\overline{V}_S - \overline{V}_R}{jX} \tag{2.42}$$

At the receiving-end, the complex power can be written as

$$S_R = P_R + jQ_R = V_R\overline{I}* \tag{2.43}$$

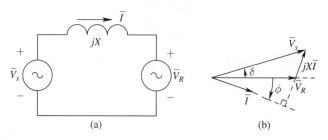

(a)　　　　　(b)

FIGURE 2.17 Power transfer between two AC systems.

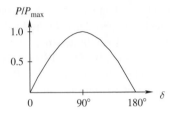

FIGURE 2.18 Power as a function of δ.

Using the complex conjugate from Equation 2.42 into Equation 2.43

$$P_R + jQ_R = V_R \left(\frac{V_S \angle (-\delta) - V_R}{-jX} \right) = \underbrace{\frac{V_S V_R \sin \delta}{X}}_{(=P_R)} + j \underbrace{\left(\frac{V_S V_R \cos \delta - V_R^2}{X} \right)}_{(=Q_R)} \qquad (2.44)$$

$$\therefore \quad P_R = \underbrace{\frac{V_S V_R}{X}}_{(=P_{max})} \sin \delta \quad \text{where} \quad P_{max} = \frac{V_S V_R}{X} \qquad (2.45)$$

which is the same as the sending end power P_S assuming no transmission-line losses. The discussion above shows that the real power P flows "downhill" on the phase angles of the voltages, and not on their magnitudes. It is plotted in Figure 2.18 for positive values of δ.

Focusing on the reactive power, from Equation 2.44,

$$Q_R = \frac{V_S V_R \cos \delta}{X} - \frac{V_R^2}{X} \qquad (2.46)$$

If the power transfer between the two systems is zero, then, from Equation 2.45, $\sin\delta$ and the angle δ are equal to zero. Under this condition, from Equation 2.46,

$$Q_R = \frac{V_S V_R}{X} - \frac{V_R^2}{X} = \frac{V_R}{X}(V_S - V_R) \qquad (\text{if } P_R = 0) \qquad (2.47)$$

which shows that the reactive power at the receiving-end is related to the difference $(V_S - V_R)$ between the two voltage magnitudes.

The real and the reactive power transfers given by Equation 2.45 through 2.47 are extremely important in the discussion of power and reactive power flows in later chapters.

2.6 APPARATUS RATINGS, BASE VALUES, AND PER-UNIT QUANTITIES

2.6.1 Ratings

All power apparatus such as generators, transformers, or transmission lines have a voltage rating V_{rated}, which is the nominal voltage that an apparatus is designed to be operated at, based on its insulation or to avoid magnetic saturation at the operating frequency, whichever may be more limiting. Similarly, each apparatus has a current rating I_{rated}, which is the nominal current, usually specified in terms of RMS, which the apparatus is designed to be operated at in steady state, and beyond which excessive $I^2 R$ heating may cause damage to the apparatus.

2.6.2 Base Values and Per-Unit Values

The parameters of an apparatus are specified in per-unit as a fraction of the appropriate base values. Often, the rated voltage and the rated current are chosen as base values and the other base quantities are calculated based on these two base values. There are several reasons for specifying apparatus parameters in per-unit:

1. Regardless of the size of the equipment, the per-unit values based on the apparatus voltage and current ratings lie in a narrow range and hence are easy to check or estimate, and
2. Power systems under study involve several transformers and hence such a system has several voltage and current ratings as we will study in the chapter on transformers. In such a system, a set of base values is chosen that is common to the entire system. Using parameters in per-unit of these common base values highly simplifies the analysis, where the voltage and current transformation due to the turns-ratio of the transformers are eliminated from calculations, as discussed in the chapter dealing with calculations of power flow.

With V_{base} and I_{base}, the other base quantities can be calculated as follows:

$$R_{base}, X_{base}, Z_{base} = \frac{V_{base}}{I_{base}} \quad (\text{in } \Omega) \tag{2.48}$$

$$G_{base}, B_{base}, Y_{base} = \frac{I_{base}}{V_{base}} \quad (\text{in } \Omega) \tag{2.49}$$

$$P_{base}, Q_{base}, (\text{VA})_{base} = V_{base} I_{base} \quad (\text{in watts, VAR, or VA}) \tag{2.50}$$

In terms of these base quantities, the per-unit quantities can be specified as

$$\text{Per-Unit Value} = \frac{\text{actual value}}{\text{base value}} \tag{2.51}$$

As mentioned earlier, the parameters of an apparatus are specified in per-unit of the base values that usually equal its rated voltage and current. Another benefit of using per-unit quantities is that their values remain the same on per-phase basis or a total three-phase basis. For example, if we are considering per-phase power, the power in watts is one-third of the total power, and so is its base value compared to that in the three-phase case.

Example 2.8
In Example 2.7, calculate the per-unit values of the per-phase voltage, the load impedance, the load current, and the load real and reactive powers, if the line-line voltage base is 208 V (RMS) and the base value of the three-phase power is 5.4 kW.

Solution   Given that $V_{LL,base} = 208$ V, $V_{ph,base} = 120$ V. The base value of the per-phase power is $P_{ph,base} = \frac{5400}{3} = 1800$ W. Therefore,

$$I_{base} = \frac{P_{ph,base}}{V_{ph,base}} = 15 \text{ A} \quad \text{and} \quad Z_{base} = \frac{V_{ph,base}}{I_{base}} = 8 \text{ }\Omega.$$

With these base values, $\overline{V}_{ph} = 1.0 \angle 0$ pu, $\overline{I} = 0.8 \angle -36.87°$ pu, $Z_L = 1.25 \angle -36.87°$ pu, $P_L = 0.64$ pu, and $Q_L = 0.48$ pu.

2.7   ENERGY EFFICIENCIES OF POWER SYSTEM APPARATUS

Power system apparatus must be energy efficient and reliable, have a high power density) (thus reducing size and weight), and be low-cost to make the overall system economically feasible. High energy efficiency is important for several reasons: it lowers operating costs by avoiding the cost of wasted energy, contributes less to global warming, and reduces the need for cooling therefore increasing power density.

The energy efficiency η of an apparatus in Figure 2.19 is

$$\eta = \frac{P_o}{P_{in}} \tag{2.52}$$

which, in terms of the output power P_o and the power loss P_{loss} within the apparatus, is

$$\eta = \frac{P_o}{P_o + P_{loss}} \tag{2.53}$$

In power systems, the apparatus such as transformers and generators have percentage energy efficiencies in upper 90s, and there is constant pursuit to increase them further.

2.8   ELECTROMAGNETIC CONCEPTS

Many of the apparatus used in power systems, like transformers, synchronous generators, transmission lines, and motor loads, require a basic understanding of electromagnetic concepts which will be reviewed in this section.

2.8.1   Ampere's Law

When a current i is passed through a conductor, a magnetic field is produced. The direction of the magnetic field depends on the direction of the current. As shown in

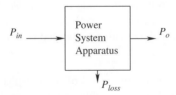

FIGURE 2.19   Energy efficiency $\eta = P_o/P_{in}$.

Figure 2.20a, the current through a conductor, perpendicular and *into* the paper plane, is represented by "×"; this current produces magnetic field in a clockwise direction. Conversely, the current *out of* the paper plane, represented by a dot, produces magnetic field in a counterclockwise direction, as shown in Figure 2.20b.

The magnetic-field intensity H produced by current-carrying conductors can be obtained by means of Ampere's Law, which in its simplest form states that, at any time, the line (contour) integral of the magnetic field intensity along *any* closed path equals the total current enclosed by this path. Therefore, in Figure 2.20c,

$$\oint H d\ell = \sum i \tag{2.54}$$

where $\oint$ represents a contour or a closed-line integration. Note that the scalar H in Equation 2.54 is the component of the magnetic field intensity (a vector field) in the direction of the differential length $d\ell$ along the closed path. Alternatively, we can express the field intensity and the differential length to be vector quantities, which will require a dot product on the left side of Equation 2.54.

Example 2.9
Consider the coil in Figure 2.21, which has $N = 25$ turns and the toroid has an inside diameter $ID = 5$ cm and an outside diameter $OD = 5.5$ cm.

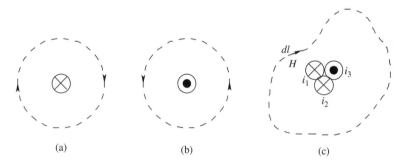

(a) (b) (c)

FIGURE 2.20 Ampere's Law.

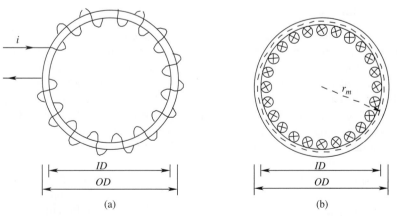

(a) (b)

FIGURE 2.21 Example 2.9.

For a current $i = 3$ A, calculate the field intensity H along the mean-path length within the toroid.

Solution   Due to symmetry, the magnetic field intensity H_m along a circular contour within the toroid is constant. In Figure 2.21, the mean radius $r_m = \frac{1}{2}(\frac{OD + ID}{2})$. Therefore, the mean path of length $\ell_m (= 2\pi r_m = 0.165$ m) encloses the current iN-times, as shown in Figure 2.21b. Therefore, from Ampere's Law in Equation 2.54, the field intensity along this mean path is

$$H_m = \frac{N\,i}{\ell_m} \tag{2.55}$$

which for the given values can be calculated as

$$H_m = \frac{25 \times 3}{0.165} = 454.5 \text{ A/m}.$$

If the width of the toroid is much smaller than the mean radius r_m, it is reasonable to assume a uniform H_m throughout the cross-section of the toroid.

The field intensity in Equation 2.55 has the units of [A/m], noting that "turns" is a unit less quantity. The product Ni is commonly referred to as the ampere-turns or mmf F that produces the magnetic field. The current in Equation 2.55 may be DC, or time-varying. If the current is time varying, the relationship in Equation 2.55 is valid on an instantaneous basis; that is, $H_m(t)$ is related to $i(t)$ by N/ℓ_m.

2.8.2   Flux Density b and the Flux ϕ

At any instant of time t for a given H-field, the density of flux lines, called the flux density B (in units of [T] for Tesla), depends on the permeability μ of the material on which this H-field is acting. In air,

$$B = \mu_o H \quad \mu_o = 4\pi \times 10^{-7} \left[\frac{\text{henries}}{\text{m}}\right] \tag{2.56}$$

where μ_o is the permeability of air or free space.

2.8.3   Ferromagnetic Materials

Ferromagnetic materials guide magnetic fields and, due to their high permeability, require small ampere-turns (a small current for a given number of turns) to produce the desired flux density. These materials exhibit the multi-valued nonlinear behavior shown by their *B-H* characteristics in Figure 2.22a. Imagine that the toroid in Figure 2.21 consists of a ferromagnetic material such as silicon steel. If the current through the coil is slowly varied in a sinusoidal manner with time, the corresponding H-field will cause one of the hysteresis loops shown in Figure 2.22a to be traced. Completing the loop once results in a net dissipation of energy within the material. This energy loss per cycle is referred as the hysteresis loss.

Increasing the peak value of the sinusoidally varying H-field will result in a bigger hysteresis loop. Joining the peaks of the hysteresis loop, we can approximate the *B-H* characteristic by the single curve shown in Figure 2.22b. At low values of magnetic field, the *B-H* characteristic is assumed to be linear with a constant slope, such that

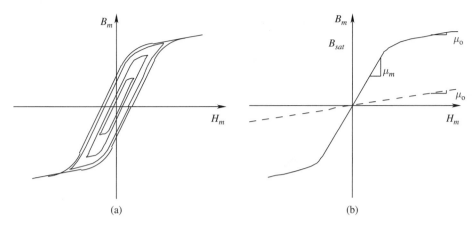

FIGURE 2.22 *B-H* characteristic of ferromagnetic materials.

$$B_m = \mu_m H_m \tag{2.57a}$$

where μ_m is the permeability of the ferromagnetic material. Typically, the μ_m of a material is expressed in terms of a permeability μ_r relative to the permeability of air:

$$\mu_m = \mu_r \mu_o \qquad \left(\mu_r = \frac{\mu_m}{\mu_o} \right) \tag{2.57b}$$

In ferromagnetic materials, μ_m can be several thousand times larger than μ_o.

In Figure 2.22b, the linear relationship (with a constant μ_m) is approximately valid until the knee of the curve is reached, beyond which the material begins to saturate. Ferromagnetic materials are often operated up to a maximum flux density, slightly above the knee of 1.6 T to 1.8 T, beyond which many more ampere-turns are required to increase flux density only slightly. In the saturated region, the incremental permeability of the magnetic material approaches μ_o, as shown by the slope of the curve in Figure 2.22b.

In this course, we will assume that the magnetic material is operating in its linear region and therefore its characteristic can be represented by $B_m = \mu_m H_m$, where μ_m remains constant.

2.8.4 Flux ϕ

Magnetic flux lines form closed paths, as shown in Figure 2.23's toroidal magnetic core, which is surrounded by the current-carrying coil.

The flux in the toroid can be calculated by selecting a circular area A_m in a plane perpendicular to the direction of the flux lines. As discussed in Example 2.9, it is reasonable to assume a uniform H_m and hence a uniform flux-density B_m throughout the core cross-section. Substituting for H_m from Equation 2.55 into Equation 2.57a,

$$B_m = \mu_m \frac{Ni}{\ell_m} \tag{2.58}$$

where B_m is the density of flux lines in the core. Therefore, making the assumption of a uniform B_m, the flux ϕ_m can be calculated as

$$\phi_m = B_m A_m \tag{2.59}$$

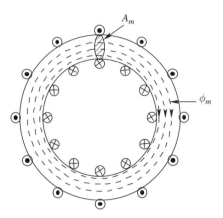

FIGURE 2.23   Toroid with flux ϕ_m.

where flux has the units of Weber [Wb]. Substituting for B_m from Equation 2.58 into Equation 2.57,

$$\phi_m = A_m \left(\mu_m \frac{Ni}{\ell_m} \right) = \frac{Ni}{\underbrace{\left(\dfrac{\ell_m}{\mu_m A_m} \right)}_{\Re_m}} \tag{2.60}$$

where Ni equals the ampere-turns (or mmf F) applied to the core, and the term within the brackets on the right side is called the reluctance $\Re_m$ of the magnetic core. From Equation 2.60,

$$\Re_m = \frac{\ell_m}{\mu_m A_m} \quad [\text{A/Wb}] \tag{2.61}$$

Equation 2.60 makes it clear that the reluctance has the units [A/Wb]. Equation 2.61 shows that the reluctance of a magnetic structure, for example the toroid in Figure 2.23, is linearly proportional to its magnetic path length and inversely proportional to both its cross-sectional area and the permeability of its material.

Equation 2.60 shows that the amount of flux produced by the applied ampere-turns $F \, (= Ni)$ is inversely proportional to the reluctance $\Re$; this relationship is analogous to Ohm's Law ($I = V/R$) in electric circuits in DC steady state.

2.8.5   Flux Linkage

If all turns of a coil, for example the one in Figure 2.23, are linked by the same flux ϕ, then the coil has a flux linkage λ, where

$$\lambda = N\phi \tag{2.62}$$

In the absence of magnetic saturation, the flux-linkage λ is related to the coil current i by the coil inductance, as illustrated in the next section.

2.8.6 Inductances

At any instant of time in the coil of Figure 2.24a, the flux linkage of the coil (due to flux lines entirely in the core) is related to the current i by a parameter defined as the inductance L_m:

$$\lambda_m = L_m i \tag{2.63}$$

where the inductance $L_m (= \lambda_m / i)$ is constant if the core material is in its linear operating region. The coil inductance in the linear magnetic region can be calculated by multiplying all the factors shown in Figure 2.24b, which are based on earlier equations:

$$L_m = \left(\frac{N}{\ell_m}\right) \mu_m A_m N = \frac{N^2}{\left(\frac{\ell_m}{\mu_m A_m}\right)} = \frac{N^2}{\mathfrak{R}_m} \tag{2.64}$$

Equation 2.64 indicates that the inductance L_m is strictly a property of the magnetic circuit (i.e., the core material, the geometry, and the number of turns), provided that the operation is in the linear range of the magnetic material, where the slope of its *B-H* characteristic can be represented by a constant μ_m.

Example 2.10
In the rectangular toroid of Figure 2.25, $w = 5$ mm, $h = 15$ mm, the mean path length $\ell_m = 18$ cm, $\mu_r = 5000$, and $N = 100$ turns. Calculate the coil inductance L_m, assuming that the core is unsaturated.

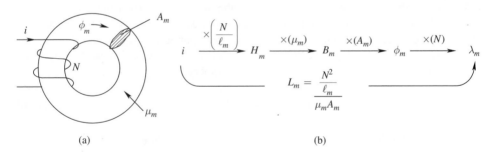

(a)   (b)

FIGURE 2.24 Coil inductance.

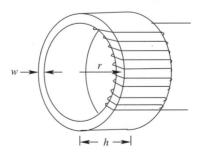

FIGURE 2.25 Rectangular toroid.

Solution   From Equation 2.61,

$$\mathfrak{R}_m = \frac{\ell_m}{\mu_m A_m} = \frac{0.18}{5000 \times 4\pi \times 10^{-7} \times 5 \times 10^{-3} \times 15 \times 10^{-3}} = 38.2 \times 10^4 \ \frac{\text{A}}{\text{Wb}}.$$

Therefore, from Equation 2.64,

$$L_m = \frac{N^2}{\mathfrak{R}_m} = 26.18 \ \text{mH}.$$

2.8.7   Faraday's Law: Induced Voltage in a Coil Due to Time-Rate of Change of Flux Linkage

In our discussion so far, we have established in magnetic circuits relationships between the electrical quantity i and the magnetic quantities H, B, ϕ, and λ. These relationships are valid under DC (static) conditions, as well as at any instant when these quantities are varying with time. We will now examine the voltage across the coil under time-*varying* conditions. In the coil of Figure 2.26, Faraday's Law dictates that the time-rate of change of flux-linkage equals the voltage across the coil at any instant:

$$e(t) = \frac{d}{dt}\lambda(t) = N\frac{d}{dt}\phi(t) \tag{2.65}$$

This assumes that all flux lines link all N-turns such that $\lambda = N\phi$. The polarity of the emf $e(t)$ and the direction of $\phi(t)$ in the above equation are yet to be justified.

The above relationship is valid, no matter what is causing the flux to change. One possibility is that a second coil is placed on the same core. When the second coil is supplied by a time-varying current, mutual coupling causes the flux ϕ through the coil shown in Figure 2.26 to change with time. The other possibility is that a voltage $e(t)$ is applied across the coil in Figure 2.26, causing the change in flux, which can be calculated by integrating both sides of Equation 2.65 with respect to time:

$$\phi(t) = \phi(0) + \frac{1}{N}\int_0^t e(\tau) \cdot d\tau \tag{2.66}$$

where $\phi(0)$ is the initial flux at $t = 0$ and τ is a variable of integration.

Recalling the Ohm's Law equation $v = Ri$, the current direction through a resistor is defined to be into the terminal chosen to be of the positive polarity. This is the passive sign convention. Similarly, in the coil of Figure 2.26, we can establish the voltage

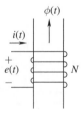

FIGURE 2.26   Voltage polarity and direction of flux and current.

polarity and the flux direction in order to apply Faraday's Law, given by Equations 2.65 and 2.66. If the flux direction is given, we can establish the voltage polarity as follows: First determine the direction of a hypothetical current that will produce flux in the same direction as given. Then, the positive polarity for the voltage is at the terminal which this hypothetical current is entering. Conversely, if the voltage polarity is given, imagine a hypothetical current entering the positive-polarity terminal. This current, based on how the coil is wound, for example in Figure 2.26, determines the flux direction for use in Equations 2.65 and 2.66. Following these rules to determine the voltage polarity and the flux direction is easier than applying Lenz's Law (not discussed here).

Another way to determine the polarity of the induced emf is to apply Lenz's Law, which states the following: if a current is allowed to flow due to the voltage induced by an increasing flux-linkage, for example, the direction of this hypothetical current will be to oppose the flux change.

Example 2.11
In the structure of Figure 2.26, the flux $\phi_m(=\hat{\phi}_m\sin\omega t)$ linking the coil is varying sinusoidally with time, where $N = 300$ turns, $f = 60$ Hz, and the cross-sectional area $A_m = 10$ cm^2. The peak flux density $\hat{B}_m = 1.5$ T. Calculate the expression for the induced voltage with the polarity shown in Figure 2.26. Plot the flux and the induced voltage as functions of time.

Solution   From Equation 2.59, $\hat{\phi}_m = \hat{B}_m A_m = 1.5 \times 10 \times 10^{-4} = 1.5 \times 10^{-3}$ Wb.
From Faraday's Law in Equation 2.63, the induced voltage below is

$$e(t) = \omega N \hat{\phi}_m \cos \omega t = 2\pi \times 60 \times 300 \times 1.5 \times 10^{-3} \times \cos \omega t = 169.65 \cos \omega t \text{ V}.$$

The induced voltage and the flux waveform are plotted in Figure 2.27, from which we can concluded that the induced voltage phasor leads the flux phasor by 90°.

Example 2.11 illustrates that the voltage is induced due to $d\phi/dt$, regardless of whether any current flows in that coil.

2.8.8   Leakage and Magnetizing Inductances

Just as conductors guide currents in electric circuits, magnetic cores guide *flux* in *magnetic circuits*. But there is an important difference. In electric circuits, the conductivity of copper is approximately 10^{20} times higher than that of air, allowing leakage currents to be neglected at DC or at low frequencies such as 60 Hz. In magnetic circuits, however, the permeabilities of magnetic materials are only around 10^4 times greater than that of air. Because of this relatively low ratio, not all the flux is confined to the core in the structure

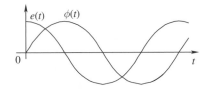

FIGURE 2.27   Example 2.11.

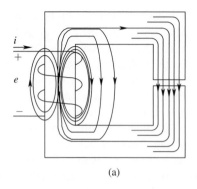
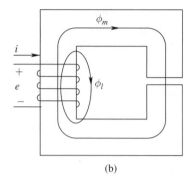

(a)   (b)

FIGURE 2.28   Including leakage flux.

of Figure 2.28a, and the core window also has flux lines which are called leakage. Note that the coil shown in Figure 2.28a is drawn schematically. In practice, the coil consists of multiple layers and the core is designed to fit as snugly to the coil as possible, thus minimizing the unused window area.

The leakage effect makes accurate analysis of magnetic circuits more difficult, so that it requires sophisticated numerical methods, such as finite element analysis. However, we can account for the effect of leakage fluxes by making certain approximations. We can divide the total flux ϕ into two parts: the magnetic flux ϕ_m, which is completely confined to the core and links all N turns, and the leakage flux, which is partially or entirely in air and is represented by an equivalent leakage flux ϕ_ℓ, which also links all N turns of the coil but does not follow the entire magnetic path, as shown in Figure 2.28b. Thus,

$$\phi = \phi_m + \phi_\ell \tag{2.67}$$

where ϕ is the equivalent flux which links all N turns. Therefore, the total flux linkage of the coil is

$$\lambda = N\phi = \underbrace{N\phi_m}_{\lambda_m} + \underbrace{N\phi_\ell}_{\lambda_\ell} = \lambda_m + \lambda_\ell \tag{2.68}$$

The total inductance (called the self-inductance) can be obtained by dividing both sides of Equation 2.68 by the current i:

$$\underbrace{\frac{\lambda}{i}}_{L_{self}} = \underbrace{\frac{\lambda_m}{i}}_{L_m} + \underbrace{\frac{\lambda_\ell}{i}}_{L_\ell} \tag{2.69}$$

Therefore,

$$L_{self} = L_m + L_\ell \tag{2.70}$$

where L_m is often called the *magnetizing inductance* due to ϕ_m in the magnetic core, and L_ℓ is called the *leakage inductance* due to the leakage flux ϕ_ℓ. From Equation 2.70, the total flux linkage of the coil can be written as

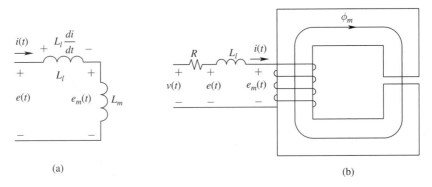

(a)   (b)

FIGURE 2.29   Analysis including the leakage flux.

$$\lambda = (L_m + L_\ell)i \tag{2.71}$$

Hence, from Faraday's Law in Equation 2.65,

$$e(t) = L_\ell \frac{di}{dt} + \underbrace{L_m \frac{di}{dt}}_{e_m(t)} \tag{2.72}$$

This results in the circuit of Figure 2.29a. In Figure 2.29b, the voltage drop due to the leakage inductance can be shown separately so that the voltage induced in the coil is solely due to the magnetizing flux. The coil resistance R can then be added in series to complete the representation of the coil.

REFERENCE

1. Any textbook in basic electric circuits and electromagnetic field theory.

PROBLEMS

2.1   Express the following voltages as phasors: (a) $v_1(t) = \sqrt{2} \times 100 \cos(\omega t - 30°)$ V and (b) $v_2(t) = \sqrt{2} \times 100 \cos(\omega t + 30°)$ V.

2.2   The series R-L-C circuit of Figure 2.3a is in a sinusoidal steady state at a frequency of 60 Hz. $V = 120$ V, $R = 1.5\ \Omega$, $L = 20$ mH, and $C = 100\ \mu$F. Calculate $i(t)$ in this circuit by solving the differential equation Equation 2.3.

2.3   Repeat Problem 2.2 using the phasor-domain analysis.

2.4   In a linear circuit in sinusoidal steady state with only one active source $\overline{V} = 90 \angle 30°$ V, the current in a branch is $\overline{I} = 5 \angle 15°$ A. Calculate the current in the same branch if the source voltage were to be $100 \angle 0°$ V.

2.5   To the circuit of Example 2.1, if a voltage $100 \angle 0°$ V is applied, calculate P, Q, and the power factor. Show that $Q = \sum_k I_k^2 X_k$.

2.6   Show that a series-connected load with R_s and X_s, as shown below in Figure P2.6a can be represented by a parallel combination, as shown in Figure 2.6b, where $R_p = (R_s^2 + X_s^2)/R_s$ and $X_p = (R_s^2 + X_s^2)/X_s$.

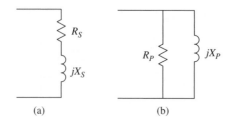

(a)   (b)

FIGURE P2.6

2.7   Using the results of the series-parallel conversion in Problem 2.6, calculate the equivalent R_P and X_P to represent the circuit of Example 2.2.

2.8   Confirm the calculations of the equivalent R_P and X_P in Problem 2.7 by using P and Q calculated in Example 2.3, recognizing that in the parallel representation, P is entirely associated with R_P, and Q entirely with X_P.

2.9   In Example 2.5, calculate the compensating capacitor in parallel, necessary to make the overall power factor to 0.9 (leading).

2.10   An inductive load connected to a 120-V (rms), 60-Hz AC source draws 5 kW at a power factor of 0.8. Calculate the capacitance required in parallel with the load in order to bring the combined power factor to 0.95 (lagging).

2.11   A positive sequence (a-b-c), balanced, wye-connected voltage source has the phase-a voltage given as $\overline{V}_a = \sqrt{2} \times 100 \angle 30°$ V. Obtain the time-domain voltages $v_a(t)$, $v_b(t)$, $v_c(t)$, $v_{ab}(t)$, and show all of these as phasors.

2.12   A balanced three-phase inductive load is supplied in steady state by a balanced, wye-connected, three-phase voltage source with a phase voltage of 120 V RMS. The load draws a total of 10 kW at a power factor of 0.9. Calculate the RMS value of the phase currents and the magnitude of the per-phase load impedance, assuming a wye-connected load. Draw a phasor diagram showing all three voltages and currents.

2.13   Repeat Problem 2.12, assuming a balanced delta-connected load.

2.14   The balanced circuit of Figure 2.14 shows the impedance of three-phase cables connecting the source terminals (a, b, c) to the load terminals (A, B, C), where $Z_{self} = (0.3 + j1.5)\ \Omega$ and $Z_{mutual} = j0.5\ \Omega$. Calculate $\overline{V}_A$ if $\overline{V}_a = 1000 \angle 0°$ V and $\overline{I}_a = 10 \angle - 30°$ A, where $\overline{V}_a$ and $\overline{V}_A$ are voltages with respect to a common neutral.

2.15   Similar to the calculation in section 2.4-2, in the balanced circuit of Figure P2.15, calculate the per-phase capacitive currents in terms of the voltages, capacitances, and frequency.

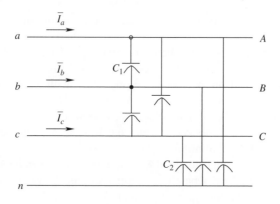

FIGURE P2.15

2.16   In the per-phase circuit of Figure 2.17a, the power transfer per-phase is 1 kW from side 1 to 2. $V_S = 100$ V, $\overline{V}_R = 95 \angle 0°$ V, and $X = 1.5$ Ω. Calculate the current, the phase angle of $\overline{V}_S$, and the per-phase Q_R supplied to the receiving end.

2.17   In a radial system represented by the circuit of Figure 2.17a, $X = 1.5$ Ω. Consider the source voltage to be constant at $\overline{V}_S = 100 \angle 0°$. Calculate and plot V_S/V_R if the load varies in a range from 0 to 1 kW at the following three power factors: unity, 0.9 (lagging), 0.9 (leading).

2.18   Repeat Example 2.8 if the base three-phase power is changed to 3.6 kW but all else remains the same.

2.19   Repeat Example 2.8 if the base line-line voltage is changed to 240 V but all else remains the same.

2.20   Repeat Example 2.8 if the base line-line voltage is changed to 240 V *and* the base three-phase power is changed to 3.6 kW.

2.21   In Example 2.9, calculate the field intensity within the core: (a) very close to the inside diameter and (b) very close to the outside diameter. (c) Compare the results with the field intensity along the mean path.

2.22   In Example 2.9, calculate the reluctance in the path of flux lines if $\mu_r = 2000$.

2.23   Consider the core of dimensions given in Example 2.9. The coil requires an inductance of $25\,\mu$H. The maximum current is 3 A and the maximum flux density is not to exceed 1.3 T. Calculate the number of turns N and the relative permeability μ_r of the magnetic material that should be used.

2.24   In Problem 2.23, assume the permeability of the magnetic material to be infinite. To satisfy the conditions of maximum flux density and the desired inductance, a small air gap is introduced. Calculate the length of this air gap (neglecting the effect of flux fringing) and the number of turns N.

APPENDIX 2A

2A.1 Line and phase currents in delta-connected load under balanced conditions

Figure 2A.1 shows a balanced delta-connected load being supplied by a balanced three-phase source. From Kirchhoff's Current Law,

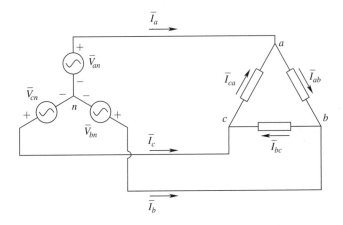

FIGURE 2A.1   Balanced delta-connected load.

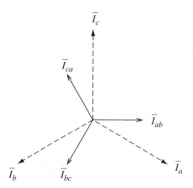

FIGURE 2A.2   Current phasors in a balanced delta-connected load.

$$\bar{I}_a = \bar{I}_{ab} - \bar{I}_{ca} \tag{2A.1}$$

Let $\bar{I}_{ab} = I_{phase} \angle\, 0°$. Since the system is balanced, $\bar{I}_{ca} = I_{phase} \angle - 240°$. Therefore, from Eq. 2A.1, as shown in Figure 2A.2,

$$\bar{I}_a = \bar{I}_{ab} - \bar{I}_{ca} = \sqrt{3}\, I_{phase} \angle - 30° \tag{2A.2}$$

Figure 2A.2 shows that the line-current magnitudes are $\sqrt{3}$ times that of the currents within the delta-connected load.

2A.2 Transformation between delta and wye-connected impedances

Consider the impedances connected in Figure 2A.3, where in general they may be unbalanced.

To arrive at the appropriate transformation, consider that the node c is disconnected from the rest of the circuit, i.e. $\bar{I}_c = 0$. Since both configurations are equivalent as far as the as external circuit is concerned, the impedance between nodes a and b must be the same. Therefore,

$$Z_a + Z_b = Z_{ab} \| (Z_{ca} + Z_{bc}) \tag{2A.3}$$

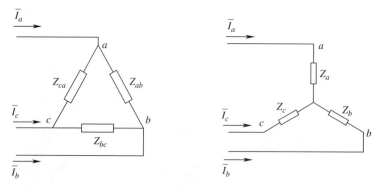

FIGURE 2A.3   Delta and wye configurations.

Similarly, considering nodes a and b to be open, respectively:

$$Z_c + Z_a = Z_{ca} || (Z_{bc} + Z_{ab}) \qquad (2A.4)$$

$$Z_b + Z_c = Z_{bc} || (Z_{ab} + Z_{ca}) \qquad (2A.5)$$

Solving the above equations simultaneously,

$$Z_a = \frac{Z_{ab}Z_{ca}}{Z_{ab} + Z_{bc} + Z_{ca}} \qquad (2A.6)$$

$$Z_b = \frac{Z_{bc}Z_{ab}}{Z_{ab} + Z_{bc} + Z_{ca}} \qquad (2A.7)$$

$$Z_c = \frac{Z_{ca}Z_{bc}}{Z_{ab} + Z_{bc} + Z_{ca}} \qquad (2A.8)$$

For the reverse transform, from Equations 2A.6 through 2A.8,

$$Z_a Z_b = \frac{Z_{ab}{}^2 Z_{bc} Z_{ca}}{(Z_{ab} + Z_{bc} + Z_{ca})^2} \qquad (2A.9)$$

$$Z_b Z_c = \frac{Z_{ab} Z_{bc}{}^2 Z_{ca}}{(Z_{ab} + Z_{bc} + Z_{ca})^2} \qquad (2A.10)$$

$$Z_c Z_a = \frac{Z_{ab} Z_{bc} Z_{ca}{}^2}{(Z_{ab} + Z_{bc} + Z_{ca})^2} \qquad (2A.11)$$

Adding the above three equations,

$$Z_a Z_b + Z_b Z_c + Z_c Z_a = \frac{Z_{ab}{}^2 Z_{bc} Z_{ca} + Z_{ab} Z_{bc}{}^2 Z_{ca} + Z_{ab} Z_{bc} Z_{ca}{}^2}{(Z_{ab} + Z_{bc} + Z_{ca})^2} \qquad (2A.12)$$

and

$$\frac{1}{Z_a}(Z_a Z_b + Z_b Z_c + Z_c Z_a) = \frac{1}{Z_a} \frac{Z_{ab}{}^2 Z_{bc} Z_{ca} + Z_{ab} Z_{bc}{}^2 Z_{ca} + Z_{ab} Z_{bc} Z_{ca}{}^2}{(Z_{ab} + Z_{bc} + Z_{ca})^2} \qquad (2A.13)$$

Substituting for Z_a from Equation 2A.6 in the right side of the above equation,

$$\frac{(Z_a Z_b + Z_b Z_c + Z_c Z_a)}{Z_a} = \frac{(Z_{ab} + Z_{bc} + Z_{ca})}{Z_{ab} Z_{ca}} \times \frac{Z_{ab}{}^2 Z_{bc} Z_{ca} + Z_{ab} Z_{bc}{}^2 Z_{ca} + Z_{ab} Z_{bc} Z_{ca}{}^2}{(Z_{ab} + Z_{bc} + Z_{ca})^2} \qquad (2A.14)$$

The right side in Equation 2A.14 simplifies to Z_{bc}, and therefore

$$Z_{bc} = \frac{(Z_a Z_b + Z_b Z_c + Z_c Z_a)}{Z_a} \qquad (2A.15)$$

By symmetry,

$$Z_{ab} = \frac{(Z_a Z_b + Z_b Z_c + Z_c Z_a)}{Z_c} \tag{2A.16}$$

and

$$Z_{ca} = \frac{(Z_a Z_b + Z_b Z_c + Z_c Z_a)}{Z_b} \tag{2A.17}$$

3

ELECTRIC ENERGY AND
THE ENVIRONMENT

3.1 INTRODUCTION

Figure 3.1a shows the production and consumption of energy by various means in the United States in 2004, which indicates that the consumption is nearly 100 quadrillion (10^{15}) BTUs where as the production is only 70 quadrillion BTUs, and the rest is being imported. (Approximately, 10,000 BTUs equal 2.93 kWh.) As shown in Figure 3.1b, out of this total energy consumption, 38.9 quadrillion BTUs, also 38.9 percent, since the consumption is nearly 100 quadrillion BTUs, are used for generating electric power which is used in various sectors, where the primary consumption figures indicate sources other than electricity. Therefore, adding all the sectors, the total consumption minus the primary consumption equals electric energy consumption.

Present installed generation capacity in the United States is approximately 10^6 MW, which is a substantial percentage of the world's total, and corresponds to roughly 3 kW per capita. Figure 3.2 shows electric power generation by various fuel sources in 2004. In the coming years and decades, energy including electric energy is going to become an increasingly precious commodity. Therefore, how we use it will be critical from the geopolitical, social, and environmental points of view. In this chapter, we will briefly examine the present and the potential choices and their environmental consequences.

3.2 CHOICES AND CONSEQUENCES

As illustrated in Figure 3.2, electric energy is derived from various sources listed below, most of them from the sun directly or indirectly, except for the nuclear energy which dates back to the creation itself.

- Hydro
- Fossil fuels: Coal, natural gas, oil
- Nuclear
- Renewable: Wind, solar, biomass, and geothermal

39

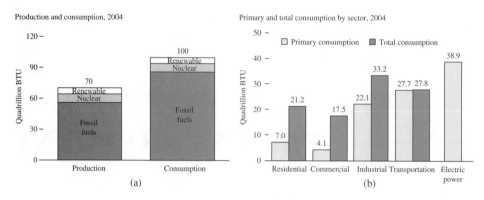

FIGURE 3.1   Production and consumption of energy in the United States in 2004 [1].

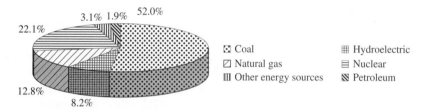

FIGURE 3.2   Electric power generation by various fuel types in the U.S. in 2005 [1].

There are environmental consequences of using energy; in fact, there is an environmental impact of all human activities. These consequences are listed below:

- Greenhouse gases, primarily carbon dioxide
- Sulfur dioxide
- Nitrogen oxides
- Mercury
- Thermal pollution

3.3   HYDRO POWER

This is one of the oldest forms for producing electricity. In hydro power plants, as shown in Figure 3.3, water in a reservoir or a dam created out of a river is discharged through penstocks into turbines that spin generators and produce electricity. In such power plants, turbine efficiencies can be better than 93 percent [2].

Hydro power can be classified based on available head (drop in the river level): high, medium, or low (run-of-river). The higher the head (H in Figure 3.3), the more potential energy can be converted by the turbine into mechanical output. Because of the environmental impact, and social implications that can be enormous, including uprooting of people, high- and medium-head hydro are generally not considered in the category of renewable energy, although technically they are. On the other hand, low-head

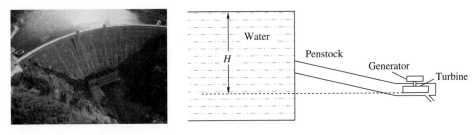

FIGURE 3.3 Hydro power (Source: www.bpa.gov).

(run-of-river) hydro, where the kinetic energy associated with flowing water is converted by a turbine into mechanical output, is considered renewable. As an example in a recent study, heads below 66 feet (20 m) were considered in the category of renewable energy. Hydro power plants can respond very quickly, in matter of a few seconds, by changing their output to meet the changes in the power demand.

3.4 FOSSIL FUEL–BASED POWER PLANTS

Fossil fuels are the main source of electric energy all over the world, except in a few countries such as hydro in Norway and nuclear in France. These fossil fuels are primarily in the following form:

• Coal
• Natural gas
• Oil

Fossil fuels are derived from sun's energy by decaying vegetation and plants. In this sense, fossil fuels are also renewable but at a time scale that is much longer—hundreds of millions of years—as compared to the human existence. All the above sources of generation are discussed briefly in the following sections.

3.4.1 Coal-Fired Power Plants

As mentioned earlier, coal-fired power plants are the main source of electricity in most countries. United States is richly endowed with this resource with 30 percent of the world's total—enough to last hundreds of years. However, there are serious environmental consequences of burning coal, namely greenhouse gases, as discussed later in this chapter. The available coal can be divided into the following categories, each of which has its own characteristics in terms of the energy content and the resulting pollution: anthracite, bituminous, sub-bituminous, lignite, and peat.

Coal is burned using various coal-firing mechanisms with different efficiencies and carbon emission into the atmosphere: mechanical stokers, pulverized-coal firing, cyclone-furnace firing, fluidized-bed combustion, and gasification. Heat from burning of coal produces steam from water, which is used in the Rankine thermodynamic cycle, shown in Figure 3.4, where water is used as a working fluid. Heat is added to the water at a high pressure in the boiler to produce steam, which expands through the turbine blades, and then is cooled in the condenser. Thermal efficiencies of such a cycle are typically in a

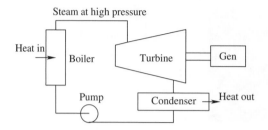

FIGURE 3.4   Rankine thermodynamic cycle in coal-fired power plants.

35–40 percent range, where for example, 9,000 BTU/kWh is typical, which corresponds to a conversion efficiency of approximately 38 percent.

In the above discussion of efficiencies, we should note that the limit of any thermal cycle efficiency is lower than that of a Carnot Cycle [3] efficiency η_c given below:

$$\eta_C = \frac{T_H - T_L}{T_H} \tag{3.1}$$

where the temperatures are in Kelvin, T_H is the higher temperature at which the heat is added, and T_L is the lower temperature at which the heat is rejected.

Coal-fired power plants have enormous environmental consequences, as discussed later in this chapter, but the availability of coal at affordable cost makes it an attractive choice. Coal plants take a long time to bring online from cold start, and changing their power output on demand, from one level to another, takes minutes.

3.4.2   Natural Gas and Oil Power Plants

Natural gas is plentiful in certain regions of the world. Gas plants do not have the same environmental consequences as the coal-fired power plants such as mercury pollution, but the natural gas being a hydrocarbon-based fuel also contributes to the greenhouse gases. They are relatively inexpensive and quick to build and can have reasonable efficiencies as discussed below. Oil power plants are similar to gas-fired power plants in their operation and efficiencies.

3.4.2.1   Single-Cycle Gas Turbines
Single-cycle gas plants use Brayton thermodynamic cycle. Brayton thermodynamic cycle can take many different forms, simplest of which is shown in Figure 3.5, where natural gas is burned in the combustion chamber in the presence of compressed air and then expanded through the turbine blades in a manner similar to steam in the Rankine cycle.

In single-cycle gas-fired power plants, efficiencies are around 35 percent that are satisfactory where primary function is to supply peak loads.

3.4.2.2   Combined-Cycle Gas Turbines
In combine-cycle gas turbines, heat in the exhaust of the Brayton cycle mentioned above is recovered to operate another steam-based Rankine cycle and the efficiency can be boosted to be in a range of 55–60 percent. Because of high efficiencies, combined-cycle power plants can be base loaded, and sizes of such plants can be as high as 500 MW.

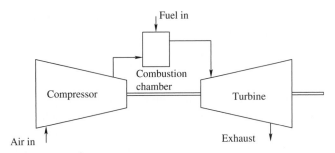

FIGURE 3.5   Brayton thermodynamic cycle in natural-gas power plants.

3.5   NUCLEAR POWER

As mentioned earlier, nuclear energy in matter dates back to the time of creation, the so called big bang, such that a gram of completely fissionable material can produce 1000 MW-Day of energy! A large nuclear plant can save 50,000 barrels of oil a day which at the present rate of above 60 dollars per barrel amounts to over 3 M$/day. Nuclear power doesn't contribute to greenhouse gases, but it creates a serious problem of storing radioactive waste for generations to come. This is a problem that has not been resolved in the United States, where there is a de facto moratorium on the construction of new nuclear power plants.

The nuclear process results in release of energy in a nuclear reactor which produces steam to spin a turbine and a generator connected to it, similar to that in coal-fired power plants using the Rankine thermodynamic cycle. Nuclear processes can be broadly classified into two categories which are briefly examined below.

3.5.1   Nuclear Fusion

Fusion is a thermonuclear reaction that is similar to the reaction that occurs incessantly on the sun where hydrogen atoms fuse together and release energy in the process. These atoms are at extremely high temperatures for this reaction to take place. In fusion reactors, two deuterium (H^2 or D) atoms fusing together result in a tritium (H^3 or T) atom, a proton (p), and in the process release 4 MeV of energy, as shown below, where 1 MeV of energy equals 4.44×10^{-20} kWh:

$$D + D \rightarrow T + p + 4 \text{ MeV} \tag{3.2}$$

Unlike fission reactors discussed in the next section, fusion reactors would have a smaller radioactive waste problem. However, the plasma that consists of fusing atoms at millions of degree in temperature must be confined away from the reactor walls, and this problem is unlikely to be solved in the near future at a commercial scale, although another attempt is now beginning to be made [4].

3.5.2   Nuclear Fission Reactors

All large commercial nuclear reactors rely on the fission process. In this process, neutrons strike atoms of fissionable material like uranium, "splitting" them, and in the process releasing more neutrons to keep this chain reaction continuing and releasing energy. This reaction can be expressed as below:

$$_{92}U^{235} + _{0}n^{1} \rightarrow _{54}Xe^{140} + _{38}Sr^{94} + 2\,_{0}n^{1} + 196 \text{ MeV} \tag{3.3}$$

where subscripts correspond to the atomic number and superscripts to the atomic mass. The reaction above shows that a neutron striking a $92U^{235}$ atom results in two other fission products, xenon $54Xe^{140}$ and strontium $38Sr^{94}$, two neutrons, and releases 196 MeV of energy. Resulting neutrons strike other uranium atoms and keep the chain reaction continuing in a controlled fashion.

In the fission process described above, the neutrons produced are at very high energy, which without being slowed would fail to strike other uranium atoms to keep the chain reaction going. Therefore, they must be "moderated," often by the coolant that transports heat and also acts as the "moderator." This chain reaction is controlled by control rods that consist of, for example, boron that absorbs neutron and thus allows control over the rate at which the chain reaction is maintained. Generally, nuclear plants are base loaded, implying that they are operated at their rated capacity. In case of emergency, nuclear reactors can be shut down by releasing chemical shim that acts as a "poison" that "kills" the nuclear reaction. However, the use of chemical shim may result in the reactor to be inoperable for days that it takes to clean away its effect.

3.5.2.1   Pressurized Water Reactors (PWR)

In the past, a few boiling-water reactors (BWR), as shown in Figure 3.6a, were built in addition to a few gas-cooled reactors (GCR). But most modern reactors are pressurized-water reactors (PWR). In PWRs, as shown in Figure 3.6b, water under pressure that is keeps it from boiling acts as a coolant to transport heat and also as a moderator to continue the chain reaction. Using a heat exchanger, the secondary cycle results in its water to boil and the steam produced is used in a Rankine thermodynamic cycle, similar to that in coal-fired plants, to spin the turbine and the generator.

3.5.2.2   Pressurized Heavy Water Reactors (PHWR)

Natural uranium is composed of less than 1 percent of U^{235} and approximately 99 percent of U^{238}. In PWR and BWR reactors where natural water is used as the moderator, enriched uranium must be used so that it has a greater percentage of U^{235} than natural uranium does. However, enriching uranium requires centrifuges which are extremely energy-expensive and difficult to justify in the absence of a large weapons program. Therefore in countries for example in Canada, CANDU reactors utilize natural uranium but use heavy water as the moderator and the coolant. Heavy water is derived from natural water in such a way it has the heavier isotope of hydrogen present. Except for this difference, PHWRs, in principle, are similar to PWRs.

3.5.2.3   Fast Breeder Reactors

These reactors utilize plutonium and produce more nuclear fuel than they consume. Therefore, they can be self-sustaining, however, their extensive use of plutonium, a substance for making nuclear weapons, makes them an extremely risky proposition, particularly in this post-9/11 world. With its large Super-Phenix reactors, France has made the largest implementation of such breeder reactors.

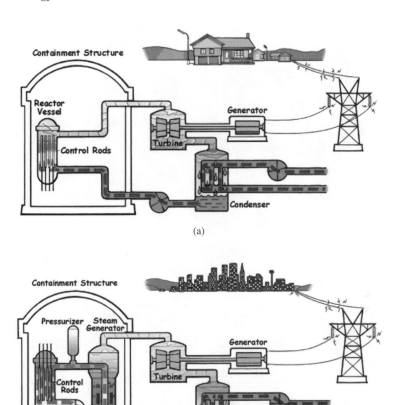

FIGURE 3.6 (a) BWR and (b) PWR reactors [5].

3.6 RENEWABLE ENERGY

Renewable energy is a means of reducing the rate of pollution caused by fossil fuel–based power plants. Renewable sources include wind, solar, biomass, and others. Some of these are described briefly.

3.6.1 Wind Energy

Wind energy is an indirect manifestation of solar energy, caused by uneven heating of the earth's surface by the sun. Out of all renewable energies, wind has come a long way and still this potential is just beginning to be realized. Figure 3.7 shows the wind potential by states in the United States, where there are several areas with good to excellent wind conditions.

In wind, the mass flow rate $\dot{m}$ in kg/s is

$$\dot{m} = \rho A V \qquad (3.4)$$

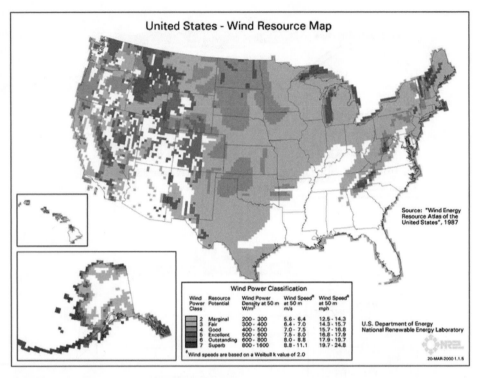

FIGURE 3.7   Wind resource map of the United States [6].

where ρ is the wind density in kg/m^3, A is the cross-sectional area in m^2 perpendicular to the wind velocity and V is the wind velocity in m/s.

Therefore, the power in wind is the rate of the kinetic energy of the wind stream:

$$P_{tot} = \frac{1}{2}\dot{m}V^2 = \frac{1}{2}\rho A V^3 \tag{3.5}$$

which shows the highly nonlinear dependence of wind power on the cube of the wind velocity. All this power cannot be removed from the wind otherwise it will "pile up" behind the turbine; an impossibility. The power that can be derived is the total power times a coefficient of performance, C_p, that is the ratio of the power available in the wind to that harnessed:

$$P_w = C_p P_{tot} = C_p \left(\frac{1}{2}\rho A V^3\right) \tag{3.6}$$

where a detailed derivation in [3] shows that at its limit, theoretically, $C_{p,\,max} = 0.5926$. As wind approaches the plane in which the turbine blades are rotating, the pressure builds up and the velocity begins to go down. After the plane of the blades, which has a very short thickness, the pressure begins to decrease and equals the pressure that exists much

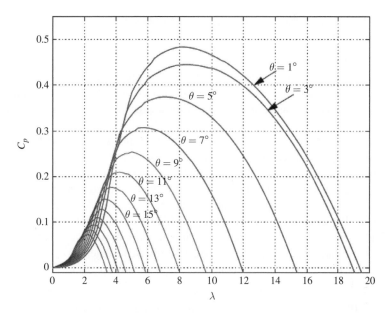

FIGURE 3.8 C_p as a function of λ [7]; these would vary based on the turbine design.

before the windmill. However, the wind-stream velocity keeps going down and levels off at a value lower than the initial velocity, due to the energy extracted from it.

This characteristic C_p in Equation 3.6 is a function of the tip-speed ratio λ, as plotted in Figure 3.8 where

$$\text{Blade Tip-Speed Ratio } \lambda = \frac{\omega_m r}{V} \tag{3.7}$$

in which r is the radius of the turbine blades in m, and ω_m is the turbine rotational speed in rad/s, and V is the wind speed in m/s.

In practice, the maximum attainable value of C_p is generally around 0.45 at a pitch-angle of $\theta = 0$ degrees (this angle is zero when the blades are vertical to the wind direction). As shown in Figure 3.8, for each blade pitch-angle θ, C_p reaches a maximum at a particular value of the tip-speed ratio λ, which, for a given wind speed V, can be obtained by controlling the turbine rotational speed ω_m. The curves for various values of the pitch-angle θ shows that the power harnessed from wind can be regulated by controlling the pitch-angle of the blades, and thus "spilling" some of the wind at very high wind speeds to prevent the power output from exceeding its design (rated) value.

3.6.1.1 Types of Generation Schemes in Windmills

Commonly used schemes for power generation in windmills require a gearing mechanism because the wind turbine rotates at very slow speeds, whereas the generator operates at a high speed close to the synchronous speed, which at the 60-Hz line frequency would be 1800 rpm for a four-pole and 900 rpm for an eight-pole machine. Therefore, the nacelle contains a gearing mechanism that boosts up the turbine speed to drive the generator at a higher speed; need for a gearing mechanism is one of the inherent drawbacks of such schemes. In very large sizes for offshore applications, there are efforts to use direct-drive (without gears) wind-turbines; however, most windmills use gearing. This subsection describes various types of wind-generation schemes.

Induction Generators, Directly Connected to the Grid. As shown in Figure 3.9, this is the simplest scheme, where a wind-turbine driven squirrel-cage induction generator is directly connected to the grid through a back-to-back connected thyristor-pair for soft-start. Therefore, it is the least expensive and uses a rugged squirrel-cage rotor induction machine.

For the induction machine to operate in its generator mode, the rotor speed must be greater than the synchronous speed. The drawback of this scheme is that since the induction machine always operates very close to the synchronous speed, it is not possible to achieve the optimum C_p at all times, and hence is not optimum at low and high wind speeds compared to variable speed schemes described below. Another disadvantage of this scheme is that a squirrel-cage induction machine always operates at a lagging power factor (that is, it draws reactive power from the grid as an inductive-load would). Therefore, a separate source, for example shunt-connected capacitors are often needed to supply the reactive power to overcome the lagging power factor operation of the induction machine.

Doubly-Fed, Wound-Rotor Induction Generators. The scheme in Figure 3.10 utilizes a wound-rotor induction machine where the stator is directly connected to the utility supply and the rotor is injected desired currents through a power-electronics interface, which is discussed in Chapter 7. Typically, four-fifths of the power flows directly from the stator to the grid and only one-fifth of the power flows through the power electronics in the rotor circuit. The drawback of this scheme is that it uses a wound-rotor induction machine where the currents to the three-phase wound rotor are supplied through slip-rings and

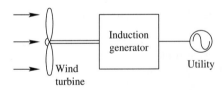

FIGURE 3.9   Induction generator directly connected to the grid [8].

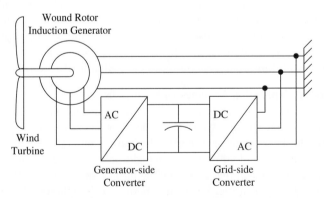

FIGURE 3.10   Doubly fed, wound-rotor induction generator [8–9].

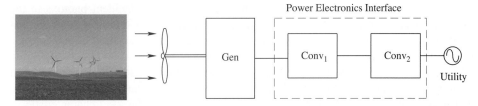

FIGURE 3.11   Power Electronics connected generator [10].

brushes which require maintenance. In spite of the fact that power electronics is expensive, since power electronics is rated only one-fifth of the system rating, therefore the overall cost is not much higher than the previous scheme. However, there are several distinct advantages over the previous scheme, as described below.

The scheme using a doubly-fed wound-rotor induction machine can typically operate in a range ±30 percent around the synchronous speed and hence it is able to capture more power at lower and higher wind speeds compared to the previous scheme because it can operate at above, as well as below, the synchronous speed. It can also supply reactive power, whereas in the previous scheme, the squirrel-cage induction machine only absorbs reactive power. Therefore, the scheme using a doubly fed wound-rotor induction machine is quite popular in windmills being installed in the United States.

Power Electronics Connected Generator. In the third scheme shown in Figure 3.11, a squirrel-cage induction generator or a permanent-magnet generator is connected to the grid through a power electronics interface described in Chapter 7. This interface consists of two converters. The converter at the generator-end supplies the reactive power excitation needed if it is an induction generator. Its frequency of operation is controlled to be optimum for the prevailing wind speed. The converter at the line-end is capable of absorbing or supplying reactive power in a continuous manner. This is the most flexible arrangement using a rugged squirrel-cage machine or a high-efficiency permanent-magnet generator, which can operate in a very wide wind-speed range and is the likely contender for the future arrangements as the cost of the power electronics interface, that must handle the entire power output of the system, is continuing to go down.

3.6.1.2   Challenges in Harnessing Wind Energy

Wind power has enormous potential. North and South Dakota alone can potentially supply two-thirds of the present electric energy needs of the United States. But wind power also has many challenges. Wind is variable and the power in it varies as the cube of the wind velocity. Therefore, it is difficult to use it as a conventional dispatchable source by energy control centers. To overcome the dispatchability problem, research is being conducted in storage for example in flywheels for a short duration, supplemented by other generation, such as biodiesel when wind dies down for longer periods. The other problem with wind resources is that they are located far away from load centers, and harnessing this energy would require new transmission lines to be built.

3.6.2   Photovoltaic Energy

As mentioned earlier, except for nuclear, all forms of energy resources are indirectly based on solar energy. Photovoltaic cells convert the energy in sun's rays directly into electricity.

Photovoltaic cells consist of a *pn* junction where due to incident photons in the sun's rays cause excess electrons and holes to be generated above their normal thermal equilibrium. This causes a potential to be developed and if the external circuit is completed, it results in flow of electrons and thus the flow of current. The $v - i$ characteristic of a photovoltaic cell is as shown in Figure 3.12, which shows that each cell produces an open-circuit voltage of approximately 0.6 V. The short-circuit current is limited and the power available is at its optimum value around the "knee" of the $v - i$ characteristic.

In PV systems, the PV arrays (typically four of them connected in series) provide a voltage of 53 V to 90 V DC, which the power electronic system, as shown in Figure 3.13, converts to 120 V/60 Hz sinusoidal voltage suitable for interfacing with the single-phase utility. A maximum-power-point tracking circuit in Figure 3.13 allows these cells to be operated at this maximum-power point.

There are several types of photovoltaic cells, like mono-crystalline silicon with efficiency of 15–18 percent, multi-crystalline silicon with efficiency of 13–18 percent, and amorphous silicon with efficiency of 5–8 percent.

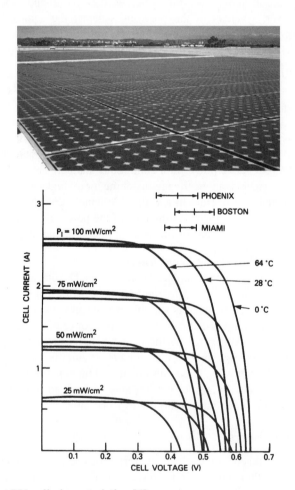

FIGURE 3.12   PV cell characteristics [6].

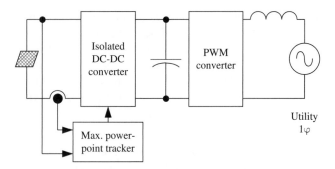

FIGURE 3.13   Photovoltaic systems.

3.6.3   Fuel Cells

Lately, there has been a great deal of interest in, and effort being devoted to, fuel cell systems. Fuel cells use hydrogen (and possibly other fuels) as the input and, through a chemical reaction, directly produce electricity with water and heat as byproducts. In fuel cells, hydrogen reacts with the catalyst to produce electrons and protons. Protons pass through a membrane. Electrons flow through the external circuit and combine with the protons and oxygen to produce water. Heat is generated in this process as a byproduct. Therefore they have none of the environmental consequences of the conventional fuels, and their efficiency can be as high as 60 percent.

Fuel-cell output is a DC voltage as shown in Figure 3.14, and therefore the need for power electronics converters to interface with the utility is the same in fuel cell systems as in photovoltaic systems.

At present, fuel cells are not commercially competitive and there are several challenges to be overcome. It is not clear what would be the best way to produce and

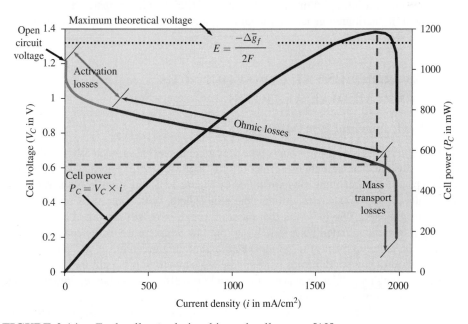

FIGURE 3.14   Fuel cell *v-i* relationship and cell power [12].

transport hydrogen. If some other fuel such as natural gas is used, what would be the economic feasibility of reformers that would be needed in conjunction with fuel cells? However, intense research is being conducted in the development of various fuel cells and some of them are listed here with expected percentage system efficiency in brackets: PEM (32–40 percent), alkaline (32–45 percent), phosphoric acid (36–45 percent), molten carbonate (43–55 percent), solid oxide (43–55 percent).

3.6.4   Biomass

There are various other sources for electrical generation under consideration. Biomass consists of biological material like agricultural waste, wood, and manure. Other possibilities are vegetable oil, ethanol from corn and other crops, and gas from anaerobic decomposition. Biomass includes refuse-derived fuel. Use of biomass and its derivatives for commercial power production is just beginning, but significant effort is being put into this field.

3.7   DISTRIBUTED GENERATION (DG)

As mentioned in Chapter 1, utility landscape is changing and although it is in its infancy, there is growing use of distributed generation. This type of generation, as the name implies, is distributed and usually much smaller in power rating than the conventional powerplants. Therefore, distributed generation is often spurred by renewable resources such as windmills. In addition, there is a movement to generate electricity local to the load so to minimize the cost of transmission and distribution lines, and the power losses associated with them. Such distributed generation may be by micro-turbines and fuel cells that may be able to utilize natural gas through a reformer. One of the significant advantages of such distributed generation would be to utilize the heat produced as a byproduct of this generation, rather than "throwing" it away as is common in central powerplants, thus resulting in much higher energy efficiency in comparison. An ultimate achievement in distributed generation would be photovoltaic if the cost of PV cells were to decrease significantly.

3.8   ENVIRONMENTAL CONSEQUENCES AND REMEDIAL ACTIONS

3.8.1   Environmental Consequences

Burning of fossil fuels in power plants produces greenhouse gases, such as CO_2, that lead to climate change. The greenhouse effect of CO_2 is illustrated in Figure 3.15. Similar to the glass panes of a greenhouse, as illustrated by Figure 3.15, radiation from the sun in the form of ultraviolet light passes through the atmosphere, with some of it absorbed and some reflected back. The radiation that passes through is mostly absorbed by the earth, warming it. Heat is radiated back, but due to the low temperature of the earth's surface, this radiation is in the form of infrared. Like the glass in a greenhouse, greenhouse gases such as carbon dioxide and nitrous oxide in the atmosphere trap the infrared heat, so less is radiated through and back to space. This greenhouse effect is leading to climate change with disastrous consequences predicted.

There are other consequences as well. Burning of fossil fuels results in increased concentration of sulfur and nitrogen oxides in the atmosphere. These get converted into

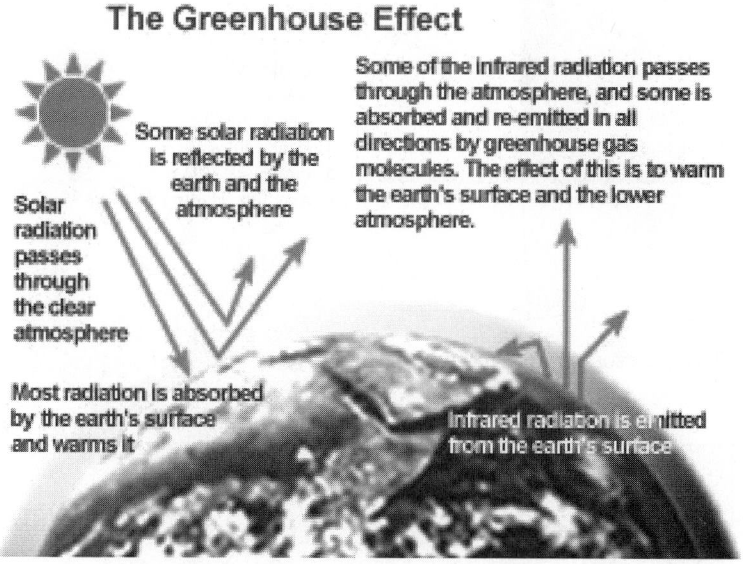

FIGURE 3.15   Greenhouse effect [13].

acid rain or acid snow with risk to health and aquatic life. Certain types of coal lead to mercury pollution that is extremely harmful to health and gets concentrated in fish, for example. Since the efficiency of thermodynamic cycle in fossil fuel plants is only around 40 percent, the rest of the energy goes directly into creating thermal pollution of the water that is used to cool it. This can have adverse consequences such as the growth of algae, etc.

3.8.2   Remedial Actions

These include flue-gas desulfurization, scrubbers for nitrogen oxide removal, and electrostatic participators. Such remedies increase power plant costs and hence of the generated electricity. Carbon sequestering is also being investigated.

3.9   RESOURCE PLANNING

Choice of using available resources for generating electricity is based on various considerations—initial investment, cost of electricity, environmental considerations, etc.—and utilities use a mix of resources, as illustrated in Figure 3.16 [14]. These include coal, nuclear, gas and oil, renewables, and purchases from other utilities.

The cost of fuel for electricity in year 2005 in the United States is shown in Figure 3.17 by months.

Some of the facts and approximate figures for year 2007 for a utility in Minnesota are as follows:

- Coal-Fired Plants: These take 6–8 years to build and cost 2,000 dollars/kW. Their thermal efficiency is in a range of 33–37 percent. The startup time associated with coal plants is 6–8 hours. Including the fuel and all other costs, the electricity cost from such

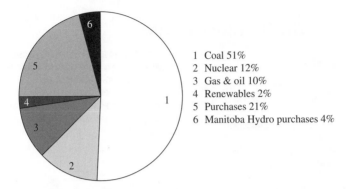

1 Coal 51%
2 Nuclear 12%
3 Gas & oil 10%
4 Renewables 2%
5 Purchases 21%
6 Manitoba Hydro purchases 4%

FIGURE 3.16   Resource mix at XcelEnergy [14].

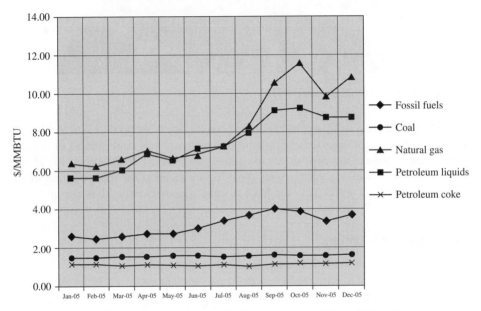

FIGURE 3.17   Electric power industry fuel costs in the U.S. in 2005 [1].

plants is 2.0–2.5 cents/kWh. (This cost is so low due to the availability of coal to this utility at cheap prices at the mine-mouth plant and the transmission of electricity over HVDC lines.)

- Single-Cycle Gas-Fired Plants: These take approximately a year to build and cost 500 dollars/kW. Their thermal efficiency is the same as the coal-fired plants, in a range of 33–37 percent. The startup time associated with gas plants is as little as 10 minutes and can be longer than 30 minutes. Including the fuel and all other costs, the electricity cost from such plants is 20 cents/kWh.

- Combined-Cycle Gas-Fired Plants: Compared to simple single-cycle gas plants their thermal efficiency can be twice as high, upwards of 60 percent. As a rule-of-thumb, if a plant is to be used for less than 20 percent of the time in a year, then the single-cycle gas turbine is selected. In a 20–60 percent range, a combined-cycle gas plant is used;

any higher than 60 percent usage favors a base-loaded coal-fired plant. Including the fuel and all other costs, the electricity cost from such plants is around 10 cents/kWh.

- Wind Turbines: Wind has become a commercial resource with the cost of electricity approximately 4.5 cents/kWh with the Federal Production Tax Credits (PTC) and 6.4 cents/kWh without them.

It should be noted that photovoltaic on the utility-scale and wave energy have the potential to become economically feasible and research into these areas is aggressively being pursued.

REFERENCES

1. U.S. Department of Energy (www.eia.doe.gov).
2. Homer M. Rustebakke (editor), *Electric Utility Systems and Practices*. 4th edition, John Wiley & Sons, 1983.
3. M. M. El-Wakil, *Powerplant Technology*, McGraw-Hill Companies, 1984.
4. *Technology Review Magazine*, MIT, September 2005.
5. http://www.nrc.gov/reading-rm/basic-ref/students/reactors.html.
6. National Renewable Energy Lab (www.nrel.gov).
7. Kara Clark, Nicholas W. Miller, Juan J. Sanchez-Gasca, *Modeling of GE Wind Turbine-Generators for Grid Studies*, GE Energy Report, Version 4.4, September 9, 2009.
8. Ned Mohan, *Electric Machines and Drives: A First Course*, John Wiley & Sons (www.wiley.com), 2011.
9. Ned Mohan, *Advanced Electric Drives* (www.mnpere.com).
10. Ned Mohan, *Power Electronics: A First Course*, John Wiley & Sons (www.wiley.com), 2011.
11. N. Mohan, *Power Electronics: Converters*, *Applications and Design*, T. Undeland, and W. P. Robbins, John Wiley & Sons (www.wiley.com).
12. www.NETL.DOE.gov.
13. http://www.epa.gov/climatechange/kids/index.html.
14. 2004 Environmental Report, www.xcelenergy.com.

PROBLEMS

3.1 What are the sources of electric energy?

3.2 What are the environmental consequences of using electric energy?

3.3 Describe hydro power plants.

3.4 Describe two nuclear processes, which one is used for commercial nuclear power production?

3.5 Describe different types of nuclear reactors.

3.6 Describe coal-fired power plants.

3.7 Describe gas-fired power plants.

3.8 What are the potential of and challenges associated with wind energy?

3.9 How does the wind-turbine efficiency depend on the tip-speed ratio and the pitch angle of the blades?

3.10 What the three most commonly used generation schemes associated with harnessing wind energy?

3.11 What is the principle underlying the operation of PV cells?

3.12   What is the principle underlying the operation of fuel cells?

3.13   What is the greenhouse effect?

3.14   What are other environmental consequences of power production from using fossil fuels?

3.15   What are the present and potential remedies to minimize environmental impacts of using fossil fuels?

3.16   A U.S. Department of Energy report estimates that over 122 billion kWh/year can be saved in the manufacturing sector in the United States by using mature and cost-effective conservation technologies. Calculate (a) how many 1000 MW generating plants are needed to operate constantly to supply this wasted energy, and (b) the annual savings in dollars if the cost of electricity is 0.10 $/kWh.

3.17   Electricity generation in the U.S. in year 2000 was approximately 3.8×10^9 MW-hrs, 16 percent of which is used for heating, ventilating and air conditioning. Based on a DOE report, as much as 30 percent of the energy can be saved in such systems by using adjustable speed drives. On this basis, calculate the savings in energy per year and relate that to 1000 MW generating plants needed to operate constantly to supply this wasted energy.

3.18   The total amount of electricity that could potentially be generated from wind in the United States has been estimated at 10.8×10^9 MW-hrs annually. If one-tenth of this potential is developed, estimate the number of 2 MW windmills that would be required, assuming that on average a windmill produces only 25 percent of the energy that it is capable of.

3.19   In problem 3.18, if each 2 MW windmill has 20 percent of its output power flowing through the power electronics interface, estimate the total rating of these power electronics interfaces in kW.

3.20   Lighting in the U.S. consumes 19 percent of the generated electricity. Compact fluorescent lamps (CFLs) consume one-fourth of the power consumed by incandescent lamps for the same light output. Electricity generation in the U.S. in year 2000 was approximately 3.8×10^9 MW-hrs. Based on this information, estimate the savings in MW-hrs annually, assuming all lighting at present is by incandescent lamps, which are going to be replaced by CFLs.

3.21   Fuel-cell systems that also utilize the heat produced can achieve efficiencies approaching 80 percent, more than double of the gas-turbine based electrical generation. Assume that 25 million households produce an average of 5 kW. Calculate the percentage of electricity generated by these fuel-cell systems compared to the annual electricity generation in the U.S. of 3.8×10^9 MW-hrs.

3.22   Induction cooking based on power electronics is estimated to be 80 percent efficient compared to 55 percent for conventional electric cooking. If an average home consumes 2 kW-hrs daily using conventional electric cooking, and that 50 million households switch to induction cooking in the U.S., calculate the annual savings in electricity usage.

3.23   Assume the average energy density of sunlight to be 800 W/m^2 and the overall photovoltaic system efficiency to be 10 percent. Calculate the land area covered with photovoltaic cells needed to produce 1,000 MW, the size of a typical large central power plant.

3.24   In problem 3.23, the solar cells are distributed on top of roofs, each in area of 50 m^2. Calculate the number of homes needed to produce the same power.

AC TRANSMISSION LINES AND UNDERGROUND CABLES

4.1 NEED FOR TRANSMISSION LINES AND CABLES

Electricity is often generated in areas that are remote to load centers, like metropolitan areas. Transmission lines form an important link in the power system structure to transport large amounts of electrical power with as little power loss as possible, keeping the system operating stably, and all at a minimum cost. Transmission line access has become an important bottleneck in power systems operation today, with building of additional transmission lines becoming a serious hurdle to overcome. This is one of the challenges facing large-scale harnessing of wind energy, for example.

Most transmission systems consist of AC overhead transmission lines. Although we will mostly discuss overhead AC transmission lines, the analysis presented in this chapter applies to underground AC cables as well, as described briefly in this chapter, which are used near metropolitan areas.

There are several high-voltage DC (HVDC) transmission systems that are also used to transport large amounts of power over long distances. Such systems are expected to be used more frequently. HVDC systems require power electronics converters and thus are discussed in a separate chapter along with other applications of power electronics in power systems.

4.2 OVERHEAD AC TRANSMISSION LINES

The discussion presented here applies to transmission as well as distribution lines, although the focus is on transmission lines. Transmission lines always consist of three phases at commonly used voltages of 115 kV, 161 kV, 230 kV, 345 kV, 500 kV and 765 kV. Voltages lower than 115 kV are generally considered for distribution. Voltages higher than 765 kV are being considered for long distance transmission.

A typical 500-kV transmission tower is shown in Figure 4.1a. There are three phases (a, b and c) where each phase consists of three conductor bundles that are hung from the steel tower through insulators, as shown in Figure 4.1b. The conductors with multiple stands of aluminum on the outside and steel at the core, as shown in Figure 4.1c, are called aluminum conductors steel reinforced (ACSR). The current through the

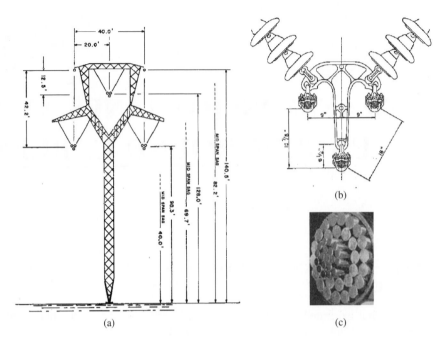

FIGURE 4.1   500-kV transmission line (Source: University of Minnesota EMTP course).

conductor primarily flows through the aluminum that has conductivity higher than seven times that of steel, which is used at the core to provide tensile strength and thus prevent these lines from excessive sagging in between the towers. Typically, there are approximately five towers per mile.

Parameters of a transmission line depend on the voltage level for which it is designed since higher voltage lines require larger values of separation between conductors, height of the conductors, and their separation from the tower that is grounded. A proper clearance is needed to keep electric field strengths at the ground level under the transmission lines to be within the limit, which is specified to be 8 kV (rms)/m by the Minnesota Environmental Quality Board, for example.

4.2.1   Shield Wires, Bundling and Cost

Shield Wires: As shown in Figure 4.1a, shield wires are often used to shield transmission-line phase conductors from being struck by lightening. These are also called ground wires, which are periodically grounded through the tower. Therefore, it is important to achieve as small a tower footing resistance as possible.

Bundling: Bundled conductors are used, as shown in Figure 4.1b, to minimize electric field strength at the conductor surface to avoid the corona effect discussed later in this chapter. According to [1], the maximum surface conductor gradient should be less than 16 kV/cm. Bundling of conductors also results in reducing the line series inductance and increasing the shunt capacitance, both of which are beneficial in loading of lines to higher power levels. However, conductor bundling results in higher cost and in increasing the clearance required from the tower [2]. At line voltages lower than 345 kV, most transmission lines consist of single conductors. At 345 kV, most transmission lines consist of bundled conductors, usually a two-conductor bundle at a spacing of 18 inches. At 500 kV, transmission lines use bundled conductors, such as a three-conductor bundle shown in Figure 4.1b, with a spacing of 18 inches.

Cost: Cost of such lines depends on a variety of factors but as a rough estimate, a 345-kV line costs range from 500,000 dollars per mile (1.6 km) in rural areas to over 2 million dollars per mile in urban areas of Minnesota, with an average of approximately 750,000 dollars per mile. Similar rough estimates are available for other transmission voltages.

4.3 TRANSPOSITION OF TRANSMISSION LINE PHASES

In the structure shown in Figure 4.1a, the phases are arranged in a triangular fashion. In other structures, these phases may be all horizontal or all vertical as shown in Figure 4.2a. It is clear from these arrangements that the electrical and magnetic couplings are not equal between the phases. For example in Figure 4.2a, coupling between phases a and b would differ that between phases a and c. To make these three phases appear balanced, they are transposed as shown in Figure 4.2b. This unbalance is not of much significance for short lines, for example less than 100 km in length. But in long lines, the length of the "barrel," equal to the three sections or one cycle of transposition as shown in Figure 4.2b, is recommended to be approximately 150 km for a triangular configuration and smaller for horizontal or vertical arrangements [3]. In spite of the fact that the transmission lines are seldom transposed, we will assume the three phases to be perfectly balanced in our simplified analysis to follow.

4.4 TRANSMISSION LINES PARAMETERS

Transmission line resistances, inductances, and capacitances are distributed all through the length of the line. Assuming the three phases to be balanced, we can easily calculate the transmission line parameters on a per-phase basis, as shown in Figure 4.3 where the bottom conductor is neutral (in general, hypothetical) that carries no current in a perfectly balanced three-phase arrangement under a balanced sinusoidal steady state operation. For unbalanced arrangements including the effect of the ground plane and the shield wires, computer programs such as PSCAD/EMTDC [4] and EMTP are available that can also include frequency dependence of the transmission line parameters.

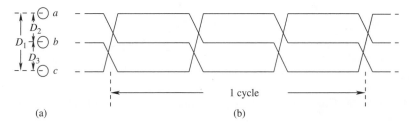

(a)   (b)

FIGURE 4.2 Transposition of transmission lines.

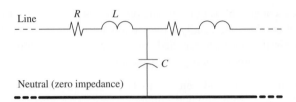

FIGURE 4.3   Distributed parameter representation on a per-phase basis.

4.4.1   Resistance *R*

The resistance of a transmission line in per unit is designed to be small to minimize I^2R power losses. These losses go down as the conductor size is increased but the costs of conductors and towers go up. In the U.S., the Department of Energy estimates that approximately 9 percent of the generated electricity is lost in transmission and distribution. Therefore, it is important to keep the transmission line resistance small. In a bundled arrangement, it is the parallel resistance of the bundled conductors that is of consideration.

There are tables [2] that give the resistance at DC and 60 Hz frequency at various temperatures. ACSR conductors such as that shown in Figure 4.1c are considered hollow as shown in Figure 4.4a because the electrical resistance of the steel core is much higher than the outer aluminum strands.

Also the skin-effect phenomenon plays a role at 60-Hz (or 50-Hz) frequency. The line resistance R depends on the length of the conductor l, the resistivity of the material ρ (that increases with temperature), and inversely on the effective cross-sectional area A of the conductor through which the current flows:

$$R = \frac{\rho l}{A} \tag{4.1}$$

The effective area A in Equation 4.1 depends on the frequency due to the skin effect, where the current at 60-Hz frequency is not uniformly distributed throughout the cross-section; rather it crowds towards the periphery of the conductor with a higher current-density J as shown in Figure 4.4b for a *solid* conductor (it will be more uniform in

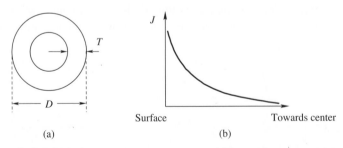

(a)   (b)

FIGURE 4.4   (a) Cross-section of ACSR conductors, (b) Skin-effect in a solid conductor.

a hollow conductor), and decreases exponentially such that at the skin-depth, the current density is a factor of $e(= 2.718)$ smaller than that at the surface. The skin depth of a material at a frequency f is [5]

$$\delta = \sqrt{\frac{2\rho}{(2\pi f)\mu}} \tag{4.2}$$

which is calculated for aluminum in the example below.

Example 4.1
In Figure 4.4a, the conductor is aluminum with resistivity $\rho = 2.65 \times 10^{-2} \mu\Omega$-m. Calculate the skin-depth δ at 60-Hz frequency.

Solution The permeability of aluminum can be consider to be that of free space, that is, $\mu = 4\pi \times 10^{-7}$ H/m. The resistivity is as specified earlier. Substituting these values in Equation 4.2,

$$\delta = 18.75 \text{ mm}$$

Therefore, in ACSR conductors, the aluminum thickness T, as shown in Figure 4.4a, is kept at the order of the skin depth and any further thickness of aluminum will essentially be a waste that will not result in decreasing the overall resistance to AC currents. In case of a type of ACSR conductor, called the Bunting conductor, shown in Figure 4.4a, $T/D = 0.3748$, where $D = 1.302$ inches. Therefore, the skin effect results in a resistance only slightly higher at 60 Hz compared to at DC in such a "hollow" conductor, where the resistance at DC is listed as 0.0787 ohms/mile versus 0.0811 ohms/mile at 60 Hz, both at the temperature of $25°C$ [2].

4.4.2 Shunt Conductance *G*

In transmission lines, in addition to the power loss due to I^2R in the series resistance, there is a small loss due to the leakage current flowing thorough the insulator. This effect is amplified due to the corona effect where the surrounding air is ionized and a hissing sound can be heard in misty, foggy weather. The corona problem can be averted by increasing the conductor size and by the use of conductor bundling as discussed earlier [2]. These losses can be represented by putting a conductance G in shunt with the capacitance in Figure 4.3, since such losses depend approximately on the square of the voltage. However, these losses are negligibly small, and thus in our analysis we will neglect the presence of G.

4.4.3 Series Inductance *L*

Figure 4.5a shows a balanced three-phase transmission line with currents i_a, i_b, and i_c, which add up to zero under balanced conditions—that is, at any time

$$i_a + i_b + i_c = 0 \tag{4.3}$$

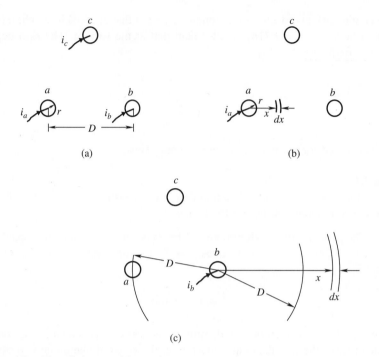

FIGURE 4.5 Flux linkage with conductor a.

We will calculate the inductance of each phase, for example phase a, as the ratio of the total flux linking it to its current. The total flux linking a phase conductor by the three currents can be obtained by superposition, and thus the per-phase inductance can be calculated as

$$L_a = \frac{\lambda_{a,total}}{i_a} = \frac{1}{i_a}\left(\lambda_{a,i_a} + \lambda_{a,i_b} + \lambda_{a,i_c}\right) \tag{4.4}$$

Considering i_a by itself, as shown in Figure 4.5b, by Ampere's Law at a distance x from conductor a,

$$H_x = \frac{i_a}{2\pi x} \quad \text{and} \quad B_x = \mu_0 H_x = \left(\frac{\mu_0}{2\pi x}\right) i_a \tag{4.5}$$

Therefore, in Figure 4.5b, the differential flux-linkage in a differential distance dx over a unit length along the conductor ($\ell = 1$) is

$$d\lambda_{x,i_a} = B_x \cdot dx = \left(\frac{\mu_0}{2\pi x}\right) i_a \cdot dx \tag{4.6}$$

Assuming the current in each conductor to be at the surface (a reasonable assumption, based on the discussion of the skin effect in calculating line resistances), integrating x from the conductor radius to infinity,

$$\lambda_{a,i_a} = \int_r^\infty d\lambda_{x,i_a} = \left(\frac{\mu_0}{2\pi}\right) i_a \int_r^\infty \frac{1}{x} \cdot dx = \left(\frac{\mu_0}{2\pi}\right) i_a \ln\frac{\infty}{r} \qquad (4.7)$$

Next, we will calculate the mutual flux linking conductor a due to i_b, as shown in Figure 4.5 c. Noting that $D \gg r$, the flux linking conductor a due to i_b is between a distance of D and infinity. Therefore, using the procedure above and using Equation 4.7,

$$\lambda_{a,i_b} = \left(\frac{\mu_0}{2\pi}\right) i_b \ln\frac{\infty}{D} \qquad (4.8)$$

Similarly, due to i_c,

$$\lambda_{a,i_c} = \left(\frac{\mu_0}{2\pi}\right) i_c \ln\frac{\infty}{D} \qquad (4.9)$$

Therefore, adding the flux linkage components from Equations 4.7 though 4.9,

$$\lambda_{a,total} = \lambda_{a,i_a} + \lambda_{a,i_b} + \lambda_{a,i_c} = \left(\frac{\mu_0}{2\pi}\right)\left[i_a\ln\frac{\infty}{r} + (i_b + i_c)\ln\frac{\infty}{D}\right] \qquad (4.10)$$

From Equation 4.3, sum of i_b and i_c equals $(-i_a)$ in Equation 4.10, which simplifies to

$$\lambda_{a,total} = \left(\frac{\mu_0}{2\pi}\right) i_a \ln\frac{D}{r} \qquad (4.11)$$

Therefore, from Equation 4.4, the inductance L associated with each phase per-unit length is

$$L = \left(\frac{\mu_0}{2\pi}\right)\ln\frac{D}{r} \quad \text{(in H/m)} \qquad (4.12)$$

If the three conductors are at distances D_1, D_2, and D_3 with respect to each other, as shown in Figure 4.2 but transposed, then the equivalent distance D (also known as the geometric mean distance *GMD*) between them, for use in Equation 4.12, can be calculated as

$$D = \sqrt[3]{D_1 D_2 D_3} \qquad (4.13)$$

In calculations above, it is assumed that the current of each conductor flows entirely at the surface. However, computer programs such as PSCAD/EMTDC can take into account the internal conductor inductance based on the current-density distribution internal to the conductor at a given frequency.

In bundled conductors, the inductance per-unit length is smaller, for example by a factor of 0.7 if a three-conductor bundle is used with a spacing of 18 inches as shown in Figure 4.1b [2]. This factor is approximately 0.8 in a two-conductor bundle with a spacing of 18 inches.

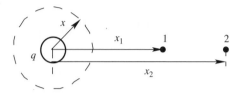

FIGURE 4.6   Electric field due to a charge.

4.4.4   Shunt Capacitance C

To calculate line capacitances, consider a charge q on a conductor, as shown in Figure 4.6, that results in dielectric flux lines and the electric field intensity E. Consider a surface at a distance x from the conductor of unit length along the conductor. The surface area per-unit length, perpendicular to the field lines, is $(2\pi x) \times 1$. Therefore, the dielectric flux-density D and the electric field E can be calculated as

$$D = \frac{q}{(2\pi x) \times 1} \quad \text{and} \quad E = \frac{D}{\varepsilon_0} = \frac{q}{(2\pi x)\,\varepsilon_0} \tag{4.14}$$

where $\varepsilon_0 = 8.85 \times 10^{-12} F/m$ is the permittivity of free space, and approximately the same for air.

Therefore, in Figure 4.6, the voltage of point 1, with respect to point 2—that is, v_{12}—is

$$v_{12} = -\int_{x_2}^{x_1} E(x)\cdot dx = \left(\frac{q}{2\pi\varepsilon_0}\right)\ln\frac{x_2}{x_1} \tag{4.15}$$

Considering three conductors at symmetrical distance D (not to be confused with the dielectric flux-density), as shown in Figure 4.7a, we can calculate voltage v_{ab}, for example, by superposing the effects of q_a, q_b, and q_c, as follows.

From Equation 4.15, where r the conductor radius, due to q_a

$$v_{ab,q_a} = \left(\frac{q_a}{2\pi\varepsilon_0}\right)\ln\frac{D}{r} \tag{4.16}$$

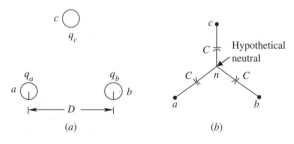

FIGURE 4.7   Shunt capacitances.

Similarly, due to q_b

$$v_{ba,q_b} = \left(\frac{q_b}{2\pi\varepsilon_0}\right)\ln\frac{D}{r} = -v_{ab,q_b} \qquad (4.17)$$

where $v_{ab,q_b} = -\left(\frac{q_b}{2\pi\varepsilon_0}\right)\ln\frac{D}{r}$.

Note that q_c, being equidistant from both the conductors a and b, does not produce any voltage between a and b—that is, $v_{ab,q_c} = 0$. Therefore, superposing the effects of all three charges

$$v_{ab} = v_{ab,q_a} + v_{ab,q_b} + \underbrace{v_{ab,q_c}}_{(=0)} = \frac{1}{2\pi\varepsilon_0}(q_a - q_b)\ln\frac{D}{r} \qquad (4.18)$$

Now consider a hypothetical neutral point n, as shown in Figure 4.7b, and the capacitances C from each phase as indicated, where

$$v_{ab} = v_{an} - v_{bn} = \frac{q_a}{C} - \frac{q_b}{C} = \frac{1}{C}(q_a - q_b) \qquad (4.19)$$

Comparing Equations 4.18 and 4.19, the shunt capacitance per-unit length is

$$C = \frac{2\pi\varepsilon_0}{\ln\frac{D}{r}} \quad (\text{in F/m}) \qquad (4.20)$$

If the three conductors are at distances D_1, D_2, and D_3, as shown in Figure 4.2, but transposed, then the equivalent distance (GMD) D between them, for use in Equation 4.20, can be calculated as

$$D = \sqrt[3]{D_1 D_2 D_3} \qquad (4.21)$$

In Equation 4.20, it should be noted that the per-phase capacitance does not include the effect of ground and the presence of shield wires. This effect is included in [4].

In bundled conductors, the per unit shunt capacitance is larger by a factor of approximately 1.4 if a three conductor bundle is used with a spacing of 18 inches [2]. This factor is approximately 1.25 in a two-conductor bundle with a spacing of 18 inches.

Typical values of transmission line parameters at various voltage levels are given in Table 4.1, where it is assumed that the conductors are arranged in three-conductor bundles with spacing of 18 inches [6].

Example 4.2

Consider a 345-kV transmission line consists of 3L3 type towers shown in Figure 4.8. This transmission system consists of a single-conductor per phase, which is a Bluebird ACSR conductor with a diameter of 1.762 inches. Ignoring the effect of ground, ground wires, and conductor sags, calculate ωL (Ω/km) and ωC ($\mu\mho$/km), and compare the results with those given in Table 4.1.

TABLE 4.1   Approximate Transmission Line Parameters
with Bundled Conductors at 60 Hz

Nominal Voltage	R (Ω/km)	ωL (Ω/km)	ωC ($\mu\mho$/km)
230 kV	0.06	0.5	3.4
345 kV	0.04	0.38	4.6
500 kV	0.03	0.33	5.3
765 kV	0.01	0.34	5.0

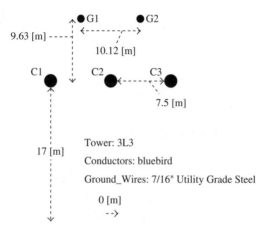

FIGURE 4.8   A 345-kV, single-conductor per phase, transmission system.

Solution   The distances between conductors in Figure 4.8 are 7.5 m, 7.5 m, and 15 m, respectively. Therefore, from Equation 4.21, the equivalent distance between them is $D = \sqrt[3]{7.5 \times 7.5 \times 15} = 9.45$ m, and the radius is $r = 0.0224$ m. Therefore at 60 Hz, from Equations 4.12 and 4.20, $\omega L = 0.456$ Ω/km and $\omega C = 3.467$ $\mu\mho$/km. These values are for a specific non-bundled conductor geometry, whereas the parameters in Table 4.1 are typical values for bundled conductors. Therefore, the values of inductances and capacitances calculated from Equations 4.12 and 4.20, modified by the factors to account for bundling mentioned earlier are in the "ballpark," whereas more accurate values are calculated using EMTDC in one of the homework problems.

4.5   DISTRIBUTED-PARAMETER REPRESENTATION OF TRANSMISSION LINES IN SINUSOIDAL STEADY STATE

A three-phase transmission line, assumed to be perfectly transposed, can be treated on a per-phase basis, as shown in Figure 4.9 under balanced operation in sinusoidal steady state.

Transmission lines are designed to have their resistance as small as it is economically feasible. Therefore, in transmission lines which are of medium length (300 km or so) or

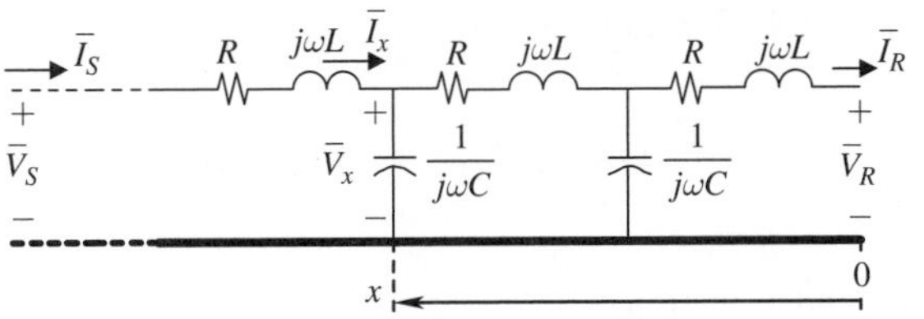

FIGURE 4.9 Distributed per-phase transmission line (G not shown).

shorter, it is reasonable to assume their resistance to be lumped, which can be taken into account separately.

With the distance x measured from the receiving end as shown in Figure 4.9 and omitting the resistance R for now, at a distance x, for a small distance Δx

$$\Delta \overline{V}_x = j\omega(L\Delta x)\overline{I}_x \quad \text{or} \quad \frac{d\overline{V}_x}{dx} = j\omega L\overline{I}_x \tag{4.22}$$

Similarly, the current flowing through the distributed shunt capacitance, for a small distance Δx

$$\Delta \overline{I}_x = j\omega(C\Delta x)\overline{V}_x \quad \text{or} \quad \frac{d\overline{I}_x}{dx} = j\omega C\overline{V}_x \tag{4.23}$$

Differentiating Equation 4.22 with respect to x and substituting into Equation 4.23,

$$\frac{d^2\overline{V}_x}{dx^2} + \beta^2\overline{V}_x = 0 \tag{4.24}$$

where

$$\beta = \omega\sqrt{LC} \tag{4.25}$$

is the propagation constant. Similarly, for the current,

$$\frac{d^2\overline{I}_x}{dx^2} + \beta^2\overline{I}_x = 0 \tag{4.26}$$

We should note that under the assumption of perfect transposition that allows us to treat a three-phase transmission line on a per-phase basis, the voltage equation of Equation 4.24 is independent of current, while the current equation of Equation 4.26 is independent of voltage. The solution of Equation 4.24 is of the following form:

$$\overline{V}_x(s) = \overline{V}_1 e^{\beta jx} + \overline{V}_2 e^{-j\beta x} \tag{4.27}$$

where $\overline{V}_1$ and $\overline{V}_2$ are coefficients that will be calculated based on the boundary conditions. The current in Equation 4.26 will have a solution of a similar form as Equation 4.27.

Differentiating Equation 4.27 with respect to x,

$$\frac{d\overline{V}_x}{dx} = j\beta\left(\overline{V}_1 e^{j\beta x} - \overline{V}_2 e^{-j\beta x}\right) \tag{4.28}$$

From Equation 4.28 and Equation 4.22,

$$\overline{I}_x = \left(\overline{V}_1 e^{j\beta x} - \overline{V}_2 e^{-j\beta x}\right)/Z_c \tag{4.29}$$

where Z_c is the surge impedance

$$Z_c = \sqrt{\frac{L}{C}} \tag{4.30}$$

Applying the boundary conditions in Equations 4.27 and 4.29—that is, at the receiving-end $x = 0$, $\overline{V}_{x=0} = \overline{V}_R$, and $\overline{I}_{x=0} = \overline{I}_R$,

$$\overline{V}_1 + \overline{V}_2 = \overline{V}_R \quad \text{and} \quad \overline{V}_1 - \overline{V}_2 = Z_c \overline{I}_R \tag{4.31}$$

From Equation 4.31,

$$\overline{V}_1 = \frac{\overline{V}_R + Z_c \overline{I}_R}{2} \quad \text{and} \quad \overline{V}_2 = \frac{\overline{V}_R - Z_c \overline{I}_R}{2} \tag{4.32}$$

Substituting from Equation 4.32 into Equation 4.27, and recognizing $\dfrac{e^{j\beta x} + e^{-j\beta x}}{2} = \cos \beta x$ and $\dfrac{e^{-j\beta x} - e^{-j\beta x}}{2} = j\sin \beta x$,

$$\overline{V}_x = \overline{V}_R \cos \beta x + j Z_c \overline{I}_R \sin \beta x \tag{4.33}$$

Similarly, from 4.32 into Equation 4.29

$$\overline{I}_x = \overline{I}_R \cos \beta x + j \frac{\overline{V}_R}{Z_c} \sin \beta x \tag{4.34}$$

4.6   SURGE IMPEDANCE Z_c AND THE SURGE IMPEDANCE LOADING (SIL)

Consider that a lossless transmission line is loaded by a resistance that equals Z_c, as shown in Figure 4.10a. Assuming the receiving end voltage to be the reference phasor,

$$\overline{V}_R = V_R \angle 0° \quad \text{and} \quad \overline{I}_R = \frac{V_R}{Z_c} \angle 0° \tag{4.35}$$

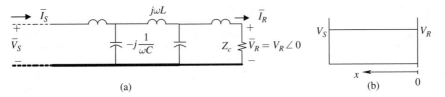

FIGURE 4.10 Per-phase transmission line terminated with a resistance equal to Z_c.

Therefore, from Equations 4.33 and 4.35, recognizing that $\overline{V}_R = V_R \angle 0°$

$$\overline{V}_x = V_R(\cos \beta x + j\sin \beta x) = V_R e^{j\beta x} \tag{4.36}$$

Similarly,

$$\overline{I}_x = I_R(\cos \beta x + j\sin \beta x) = I_R e^{j\beta x} \tag{4.37}$$

The voltage equation given by Equation 4.36 shows that the voltage magnitude has a flat profile, as shown in Figure 4.10b—that is, it is the same magnitude everywhere on the line, only the angle increases with distance x. This loading is called the surge impedance loading (SIL); the reactive power consumed by the line everywhere is the same as the reactive power produced:

$$\omega L I_x^2 = V_x^2 \omega C \tag{4.38}$$

which is true, since under SIL, from Equations 4.36 and 4.37, $\overline{V}_x / \overline{I}_x = Z_c$.

For a given voltage level, the characteristic impedance of a transmission line falls in a narrow range because the separation and the height of the conductors above ground depend on this voltage level. Table 4.2 shows the typical values for the surge impedance and the three-phase surge impedance loading (SIL), where

$$SIL = \frac{V_{LL}^2}{Z_c} \tag{4.39}$$

From Table 4.2, it is clear that for a large transfer of power, for example 1,000 MW, calls for a high voltage line such as at 345 kV or 500 kV; otherwise many circuits in parallel will be needed.

TABLE 4.2 Approximate Surge Impedance
and Three-Phase Surge Impedance Loading

Nominal Voltage	Z_c (Ω)	*SIL* (MW)
230 kV	385	135 MW
345 kV	275	430 MW
500 kV	245	1020 MW
765 kV	255	2300 MW

TABLE 4.3   Approximate Loadability of Transmission Lines

Line Length (km)	Limiting Factor	Multiple of SIL
<100	Thermal	>3
100−300	5% Voltage Drop	1.5−3
>300	Stability	1.0−1.5

Example 4.3

For the 345-kV transmission line described in Example 4.2, calculate the surge imped-ance Z_c and the Surge Impedance Loading *SIL*. If the line is 100 km long and the line resistance is 0.031 Ω/km, calculate the percentage loss if the line is surge-impedance loaded.

Solution   In the single-conductor transmission line of Example 4.2, we calculated that $\omega L = 0.456\ \Omega$/km and $\omega C = 3.467\ \mu\mho$/km. Therefore, from Equation 4.30, $Z_c \simeq 363\ \Omega$ and from Equation 4.39, *SIL* = 328 MW. At surge-impedance loading of this transmis-sion line, the per-phase current through the line is

$$I = \frac{345 \times 10^3/\sqrt{3}}{Z_c} = 548.7\ \text{A}$$

and therefore the power loss as a percentage of *SIL* in this 100 km long line, is

$$\%\ I^2 R = 100 \times \frac{3 \times 548.7^2 \times (0.031 \times 100)}{328 \times 10^6} \simeq 0.85\%$$

4.6.1   Line Loadability [2, 6, 7]

The Surge Impedance Loading (*SIL*) provides a benchmark in terms of which the amount of maximum loading of a transmission line can be expressed. This loading is a function of the length of the transmission line so that certain constraints are met. This approximate loadability is a function of the line length. Short lines less than 100 km in length can be loaded more than three times SIL, based on not exceeding the thermal limits. Medium length lines, in length (100−300 km) can be loaded to 1.5 to 3 times the SIL before the voltage drop across them exceeds more than 5 percent. Long lines above 300 km can be loaded to around the SIL because of the stability limit (discussed later in this book) so that the phase angle of the voltage between the two ends does not exceed 40 to 45 degrees. The loadability of medium and long length lines can be increased by providing series and shunt compensation discussed in the chapter dealing with stability. These results are summarized in Table 4.3.

4.7   LUMPED TRANSMISSION LINE MODELS IN STEADY STATE

In a balanced sinusoidal steady state, it is very useful to have per-phase model of transmission lines on the assumption that the three phases are perfectly transposed and the applied voltages and currents are balanced and are in a sinusoidal steady state.

In the distributed-parameter representation of the transmission line in the previous section, at the sending end with $x = \ell$,

$$\overline{V}_S = \overline{V}_R \cos \beta\ell + jZ_c \overline{I}_R \sin \beta\ell \tag{4.40}$$

and

$$\overline{I}_S = j \frac{\overline{V}_R}{2} \sin \beta\ell + \overline{I}_R \cos \beta\ell \tag{4.41}$$

This transmission line can be represented by a two-port shown in Figure 4.11, where the symmetry is necessary from both ports due to the bilateral nature of the transmission line. To find Z_{series} and $Y_{shunt}/2$, the following procedure is followed: hypothetically short-circuiting the receiving end results in $\overline{V}_R = 0$, and from Figure 4.11 and Equation 4.43,

$$\left. \frac{\overline{V}_S}{\overline{I}_R} \right|_{\overline{V}_R=0} = Z_{series} = jZ_c \sin \beta\ell \tag{4.42}$$

For transmission lines of medium length or shorter, $\beta\ell$ is small (see the homework problem). Recognizing that for small values of $\beta\ell$, $\sin \beta\ell \simeq \beta\ell$. Making use of this approximation in Equation 4.42, and recognizing that $\beta = \omega\sqrt{LC}$ and $Z_c = \sqrt{L/C}$,

$$Z_{series} = j\omega L_{line} \quad \text{where} \quad L_{line} = \ell L \tag{4.43}$$

With the receiving end open-circuited in Figure 4.11, $\overline{I}_R = 0$, and from Equation 4.40,

$$\left. \frac{\overline{V}_S}{\overline{V}_R} \right|_{\overline{I}_R=0} = 1 + Z_{series} \frac{Y_{shunt}}{2} = \cos \beta\ell \tag{4.44}$$

Therefore, from Equations 4.43 and 4.44,

$$\left(\frac{Y_{shunt}}{2} \right) = \frac{\cos \beta\ell - 1}{j\omega L_{line}} \tag{4.45}$$

From Taylor Series expansion for small values of $\beta\ell$, $\cos \beta\ell \simeq 1 - \frac{(\beta\ell)^2}{2}$ and therefore in Equation 4.45,

$$\frac{Y_{shunt}}{2} = j \frac{\omega C_{line}}{2} \quad \text{where} \quad C_{line} = C\ell \tag{4.46}$$

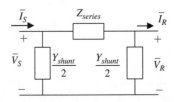

FIGURE 4.11 Lumped representation.

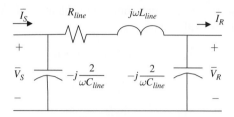

FIGURE 4.12   Per-phase representation for medium-length transmission lines.

Therefore, the per-phase equivalent circuit representation in Figure 4.11 becomes as shown in Figure 4.12, where the series resistance is explicitly shown as a lumped element. This can be confirmed from the expressions derived in the appendix for long-length lines, where the line resistance R is also considered as distributed, and simplified for medium-length lines (see the homework problem).

4.7.1   Short-Length Lines

In short lines less than 100 km in length, the effect of capacitive vars is small compared to the system strength to provide Q and can often be neglected. Therefore, the shunt capacitances shown in Figure 4.12 can be neglected. This results in only the series impedance where the series resistance may even be neglected in some simplified studies.

4.7.2   Long-Length Lines

As derived in the Appendix A, for long lines exceeding 300 km in length, it would be prudent to use a more exact representation where the parameters in the per-phase equivalent circuit of Figure 4.11 are as follows:

$$Z_{series} = Z_c \sinh \gamma \ell \tag{4.47}$$

$$\frac{Y_{shunt}}{2} = \frac{\tanh\left(\frac{\gamma \ell}{2}\right)}{Z_c} \tag{4.48}$$

where

$$Z_c = \sqrt{\frac{R + j\omega L}{G + j\omega C}} \qquad \text{and} \tag{4.49}$$

$$\gamma = \sqrt{(R + j\omega L)(G + j\omega C)} \tag{4.50}$$

4.8   CABLES [8]

As stated in [8], the underground transmission cable usage in the United States is very small: less than 1 percent of overhead line mileage. The highest underground cable voltage that is commonly used in the United States is 345 kV, and a large portion of this cable system is high-pressure fluid-filled pipe-type cable. Extruded dielectric cables are commonly used in the United States up to 230 kV, with up to 500 kV in service overseas. Underground transmission cable is generally more expensive than overhead lines. Because of all the variables (system design, route considerations, cable type, raceway type, etc.), it has to be determined case by case if underground transmission cable is a viable alternative. A rule of thumb is that underground transmission cable will cost from three to twenty times the cost of overhead line construction. As a result of the high cost, the use of high-voltage power cable for transmission and sub-transmission is generally limited to special applications caused by environmental and/or right-of-way restrictions. If underground transmission cable is going to be considered, an engineering study is required to properly evaluate the possible underground alternatives.

Underground cables are however used for power transfer in and around metro-politan areas due to lack of space for overhead transmission lines and for aesthetic reasons. Cables are also more reliable, not exposed to the elements of nature like the overhead transmission lines. For long-distance bulk power transmission, cables are generally not used because of their cost compared to the overhead transmission lines, although new trenching technologies and cable materials may reduce this cost disadvantage.

Underground cables have much larger capacitance than the overhead transmission lines and hence their characteristic impedance Z_c is much smaller. However, in spite of lower values of Z_c and higher values of *SIL*, loading of cables is limited by the problem of getting rid of the dissipated heat. For underwater transmission, cables are used in DC systems for the following reason: because of the high capacitance of underground cables, operating at 60/50-Hz AC would require shunt reactors at periodic distances to com-pensate the capacitive charging currents. Such underwater compensation reactors are not feasible.

There are various types of cables in use: pipe-type oil filled cables, self-contained oil-filled cables with a single core, cross-linked cables with polyethylene insulation, and gas-insulated cables with compressed SF_6 gas [9]. Research is being done on super-conducting cables as well. The modeling of cable parameters for power system studies is similar to those for overhead transmission lines.

REFERENCES

1. United States Department of Agriculture, Design Manual for High Voltage Transmission Lines, RUS BULLETIN 1724E-200.
2. Electric Power Research Institute (EPRI), *Transmission Line Reference Book*: 345 kV *and above*, 2nd edition.
3. Hermann W. Dommel, *EMTP Theory Book*, BPA, August 1986.
4. PSCAD/EMTDC, Manitoba HVDC Research Centre: www.hvdc.ca.
5. N. Mohan, T. Undeland and W.P. Robbins, Power Electronics: Converters, Applications, and Design, 3rd edition, *Year* 2003, John Wiley & Sons.
6. Prabha Kundur, *Power System Stability and Control*, McGraw Hill, 1994.
7. R. Dunlop et al., "Analytical Development of Loadability Characteristics for EHV and UHV Transmission Lines," IEEE Transaction on PAS, Vol PAS-98, No.2, March/April 1979.
8. United States Department of Agriculture, Rural Utilities Service, Design Guide for Rural Substations, RUS BULLETIN 1724E-300 (http://www.rurdev.usda.gov/RDU_Bulletins_-Electric.html).
9. EPRI's *Underground Transmission Systems Reference Book*.

PROBLEMS

4.1   According to [1], the conductor surface gradient in a single-conductor per-phase
transmission line can be calculated as follows:

$$g = \frac{kV_{LL}}{\sqrt{3}\, r\ln\frac{D}{r}}$$

where kV_{LL} is line-line voltage in kV, r is the conductor radius in cm, D is
the geometric mean distance (GMD) of the phase conductors in cm, and g is the
conductor surface gradient in kV/cm.

Calculate the conductor surface gradient for a 230-kV line (a) with Dove
ACSR conductors for which $r = 1.18$ cm, and (b) with Pheasant ACSR conductors for which $r = 1.755$ cm. Assume that $D = 784.9$ cm in both cases.

4.2   According to [1], the conductor surface gradient in two-conductor bundle per-
phase transmission line can be calculated as follows:

$$g = \frac{kV_{LL}(1 + 2r/s)}{2\sqrt{3}\, r\ln\frac{D}{\sqrt{rs}}}$$

where kV_{LL} is line-line voltage in kV, r is the conductor radius in cm, D is the
geometric mean distance (GMD) of the phase conductors in cm, s is the separation
between subconductors in cm, and g is the conductor surface gradient in kV/cm.

Calculate the conductor surface gradient for a 345-kV line with a two-
conductor bundle per phase with Drake ACSR conductors for which
$r = 1.407$ cm. Assume that $D = 914$ cm and $s = 45.72$ cm (18 inches).

4.3   The parameters for a 500-kV transmission system shown in Figure 4.1 with
bundled conductors are as follows: $Z_c = 258\ \Omega$ and $R = 1.76 \times 10^{-2}\ \Omega/\text{km}$.
Calculate the value for the Surge Impedance Loading *SIL* and the percentage
power loss in this transmission line if it is 300 km long and is loaded to its *SIL*.

4.4   The 500-kV line of the type in Problem 4.3 is a short-length line, 80 km long. It is
loaded to three times its *SIL*. Calculate the power loss in this line as a percentage
of its loading.

4.5   The 345-kV transmission line of the type described in Example 4.3 is 300-km
long. Calculate the parameters for its model as a long-length line. Compare these
parameter values for the model that assumes it to be a medium-length line.
Compare the percentage errors in assuming it to be a medium-length line.

4.6   Consider a 345-kV transmission line that has parameters similar to those described in
Table 4.1. Assuming a 100 MVA base, obtain its parameters in per unit.

4.7   Consider a 300-km long, 345-kV transmission line that has parameters similar to
those described in Table 4.1. Assume its receiving-end voltage $V_R = 1.0$ pu. Plot
the voltage ratio V_s/V_R as a function of P_R/SIL, where P_R is the unity power
factor load at the receiving-end. P_R/SIL varies in a range from 0 to 3.

4.8   Repeat Problem 4.7 if the load impedance has a power factor of (a) 0.9 lagging,
and (b) 0.9 leading. Compare these results with a unity-power-factor loading.

4.9   Calculate the phase angle difference of voltages between the two ends of the line
in Problem 4.7 that has unity-power-factor loading.

4.10 A 200 km long 345-kV line has the parameters given in Table 4.1. Neglect the
resistance. Calculate the voltage profile along this line if it's loaded to (a) 1.5

times *SIL*, and (b) 0.75 times *SIL*, if both ends are held at the voltages of 1 per unit.

4.11 In Problem 4.10, calculate the reactive power at both ends under the two levels of loading.

4.12 In a 230-kV cable system, the surge impedance (ignoring losses) is 25 ohms and the charging is 14.5 MVA/km. The cable length is 20 km. The resistance is 0.05 ohms/km. Both ends of this cable system are held at 1 per unit, with the sending-end voltage $\overline{V}_s = 1.0 \angle 10°$ pu and the receiving-end voltage $\overline{V}_r = 1.0 \angle 0°$ pu. Calculate the power loss in the cable, and express it as a percentage of the power received at the receiving-end.

4.13 Show that the expressions derived in the appendix for long-length lines, where the line resistance R is also considered as distributed, simplify as shown in Figure 4.12 for medium-length lines.

4.14 Show that for a 345-kV medium-length line of 200 km, $\beta\ell$ is small such that the approximations $\sin \beta\ell \simeq \beta\ell$ and $\cos \beta\ell \simeq 1 - \frac{(\beta\ell)^2}{2}$ are reasonable.

EMTDC-BASED PROBLEMS

4.15 Calculate the parameters for the single-conductor per phase, 345-kV transmission line in Example 4.2, using the data file for this problem on the accompanying website.

4.16 Calculate the parameters for the two-conductor bundle per phase, 345-kV transmission line using the data file for this problem on the accompanying website.

4.17 Calculate the parameters for the three-conductor bundle per phase, 500-kV transmission line using the data file for this problem on the accompanying website.

4.18 Obtain the results of Problem 4.11 in EMTDC using the data file for this problem on the accompanying website.

APPENDIX 4A LONG TRANSMISSION LINES

In this appendix, we will consider transmission lines of length exceeding 300 km. The procedure for determining their representation is similar to that for the medium-length lines, except the line resistance R is also considered as distributed, and the distributed shunt conductance G is also included.

With the distance x measured from the receiving end as shown in Figure 4.9, at a distance x

$$\frac{d\overline{V}_x}{dx} = (j\omega L + R)\overline{I}_x \tag{A4.1}$$

Similarly, due to the current flowing through distributed shunt capacitance,

$$\frac{d\overline{I}_x}{dx} = (j\omega C + G)\overline{V}_x \tag{A4.2}$$

Differentiating Equation A4.1 with respect to x and substituting into Equation A4.2,

$$\frac{d^2\overline{V}_x}{dx^2} = (j\omega L + R)(j\omega C + G)\overline{V}_x = \gamma^2\overline{V}_x \tag{A4.3}$$

where

$$\gamma = \sqrt{(R+j\omega L)(G+j\omega C)} = \alpha + j\beta \tag{A4.4}$$

is the propagation constant where α and β are positive in values. Similarly, for the current,

$$\frac{d^2\bar{I}_x}{dx^2} = (R+j\omega L)(G+j\omega C)\bar{I}_x = \gamma^2\bar{I}_x \tag{A4.5}$$

The solution of Equation A4.3 is of the following form:

$$\bar{V}_x = \bar{V}_1 e^{\gamma x} + \bar{V}_2 e^{-\gamma x} \tag{A4.6}$$

where $\bar{V}_1$ and $\bar{V}_2$ are coefficients that will be calculated based on the boundary conditions. The current in Equation A4.5 will have a solution of a similar form as Equation A4.6. Differentiating Equation A4.6 with respect to x,

$$\frac{d\bar{V}_x}{dx} = \gamma\left(\bar{V}_1 e^{\gamma x} - \bar{V}_2 e^{-\gamma x}\right) \tag{A4.7}$$

Comparing Equation A4.7 with Equation A4.1,

$$\bar{I}_x = \left(\bar{V}_1 e^{\gamma x} - \bar{V}_2 e^{-\gamma x}\right)/Z_c \tag{A4.8}$$

where Z_c is the surge impedance

$$Z_c = \sqrt{\frac{R+j\omega L}{G+j\omega C}} \tag{A4.9}$$

Applying the boundary conditions in Equations A4.6 and A4.8—that is, at the receiving-end $x = 0$, $\bar{V}_{x=0} = \bar{V}_R$, and $\bar{I}_{x=0} = \bar{I}_R$

$$\bar{V}_1 + \bar{V}_2 = \bar{V}_R \quad \text{and} \quad \bar{V}_1 - \bar{V}_2 = Z_c\bar{I}_R \tag{A4.10}$$

From Equation A4.10,

$$\bar{V}_1 = \frac{\bar{V}_R + Z_c\bar{I}_R}{2} \quad \text{and} \quad \bar{V}_2 = \frac{\bar{V}_R - Z_c\bar{I}_R}{2} \tag{A4.11}$$

Substituting from Equation A4.11 and that recognizing $\dfrac{e^{\gamma x}+e^{-\gamma x}}{2} = \cosh\gamma x$ and $\dfrac{e^{\gamma x}-e^{-\gamma x}}{2} = \sinh\gamma x$,

$$\bar{V}_x = \bar{V}_R\cosh\gamma x + Z_c\bar{I}_R\sinh\gamma x \tag{A4.12}$$

Similarly,

$$\bar{I}_x = \frac{\overline{V}_R}{Z_c}\sinh\gamma x + \bar{I}_R\cosh\gamma x \tag{A4.13}$$

4A.1 Lumped Transmission Line Model in Steady State

In the distributed-parameter representation of the transmission line in the previous section at the sending end with $x = \ell$,

$$\overline{V}_S = \overline{V}_R\cosh\gamma\ell + Z_c\bar{I}_R\sinh\gamma\ell \tag{A4.14}$$

and

$$\bar{I}_S = \frac{\overline{V}_R}{2}\sinh\gamma\ell + \bar{I}_R\cosh\gamma\ell \tag{A4.15}$$

This transmission line can be represented by a two-port shown in Figure 4.11, where the symmetry is necessary from both ports due to the bilateral nature of the transmission line. To find Z_{series} and $Y_{shunt}/2$, the following procedure is followed: short-circuiting the receiving end results in $\overline{V}_R = 0$, and from Figure 4.11,

$$\frac{\overline{V}_S}{\bar{I}_R}\bigg|_{\overline{V}_R=0} = Z_{series} = Z_c\sinh\gamma\ell \tag{A4.16}$$

With the receiving end open-circuited in Figure 4.11, $\bar{I}_R = 0$,

$$\frac{\overline{V}_S}{\overline{V}_R}\bigg|_{\bar{I}_R=0} = 1 + Z_{series}\frac{Y_{shunt}}{2} = \cosh\gamma\ell \tag{A4.17}$$

Therefore,

$$\frac{Y_{shunt}}{2} = \frac{1}{Z_c}\left(\frac{\cosh\gamma\ell - 1}{\sinh\gamma\ell}\right) \tag{A4.18}$$

From mathematical tables, for some arbitrary parameter A,

$$\frac{\cosh A - 1}{\sinh A} = \tanh\left(\frac{A}{2}\right) \tag{A4.19}$$

and therefore

$$\frac{Y_{shunt}}{2} = \frac{\tanh\left(\frac{\gamma\ell}{2}\right)}{Z_c} \tag{A4.20}$$

5

POWER FLOW IN POWER SYSTEM NETWORKS

5.1 INTRODUCTION

For planning purposes, it is important to know the power transfer capability of transmission lines to meet the anticipated load demand. It is also important to know the levels of power flow though various transmission lines under normal as well as under certain contingency outage conditions to maintain the continuity of service. Knowledge of power flows and voltage levels under normal operating conditions are also necessary in order to determine fault currents, if a line were to short-circuit, for example, and the ensuing consequences on the transient stability of the system.

A power flow program which determines this power flow is commonly used by all power companies for planning and operation purposes. These power flow calculations are usually performed on the bulk power system, where the effect of the secondary underlying network is implicitly included.

Determining power flow requires measurements of certain power system conditions. Theoretically, if all the bus voltages in terms of their magnitudes and phase angles could be measured with confidence, then the power flow calculations could simply be carried out by solving the linear circuit in which voltages and branch impedances, including the load impedances, are all given, and there would be no need for the procedure outlined in this chapter.

However, utilities measure a combination of quantities such as voltage magnitude V, real power P, and the reactive power Q at various buses. (Many fault protection relays make these measurements anyway to perform their function, and this information gathered by them is utilized for power flow calculations.) The measured information is telemetered to a central operating station, often through dedicated microwave transmitters and receivers. This information has to be instantaneous—that is, measured simultaneously in time. The system to acquire these measurements and to telemeter them to a control center is called SCADA (supervisory control and data acquisition) system. It is recognized that all the measurement transducers do not provide accurate information all the time. To overcome this problem, the system is over measured by measuring more quantities than are required. Then, a state estimator is used to throw away the "bad" and the redundant measurements on a probabilistic manner.

78

In this chapter, we will study the basic formulation of power-flow problem and discuss the most commonly used numerical solution methods to carry it out. To simplify our discussion, we will assume a balanced three-phase system, and therefore only a per-phase representation is necessary.

Next, we will describe how the system is represented for this study. The transmission lines are represented by their pi-circuit modal with a series impedance $Z_{series}(= R + j\omega L)$ and a susceptance $B_{shunt}/2$ $(= \omega C/2)$ at each end of the line. These are expressed in per unit at a common MVA and kV base. It is customary to use a three-phase 100 MVA base. Similarly, transformers are represented by their total leakage impedance expressed in per unit in terms of the common MVA base, ignoring their exciting currents. Transformers are assumed to be at their nominal turns ratio, hence the turns ratios do not enter into per-unit based calculations. Loads can be represented in a combination of different ways: by specifying their real and reactive powers (P and Q), as a constant-current source, or by their impedance that is treated as constant.

5.2 DESCRIPTION OF THE POWER SYSTEM

A power system can be considered as consisting of the following buses that are interconnected through transmission lines:

1. Load Buses where P and Q are specified. These are called *PQ* buses.
2. Generator buses where the voltage magnitude V and the power P are specified. These are called *PV* buses. If the upper and/or lower limits on the reactive power Q on a *PV* bus are specified and this limit is reached, then such a bus is treated as a *PQ* bus where the reactive power is specified at the limit that is reached.
3. A slack bus, which is essentially an "infinite" bus, where the voltage magnitude V is specified (normally 1 pu) and its phase angle is assumed to be zero as a reference angle. At this bus, P can be what it needs to be, based on the line losses, and hence it is called the slack bus which takes up the slack. Similarly, Q at this bus can be what it needs to be to hold the voltage at the specified value.
4. There are buses where there is no P and Q injections specified, and the voltage is also not specified. Often, these become necessary for including transformers. These can be considered as a subset of PQ buses with specified injections of $P = 0$ and $Q = 0$.

5.3 EXAMPLE POWER SYSTEM

To illustrate power flow calculations, an extremely simple power system consisting of three buses is shown in Figure 5.1. These three buses are connected through three 345-kV transmission lines 200 km, 150 km, and 150 km long, as shown in Figure 5.1. Similar to the values listed in Table 4-1 of Chapter 4, assume that these transmission lines with bundled conductors have the series reactance of 0.376 Ω/km at 60 Hz and the series resistance of 0.037 Ω/km. The shunt susceptance $B(= \omega C)$ is 4.5 $\mu\mho$/km.

To convert the actual quantities to per-unit, the voltage base is 345 kV (L-L). Following the convention, a common three-phase 100 MVA base is chosen. Therefore, the base impedance is

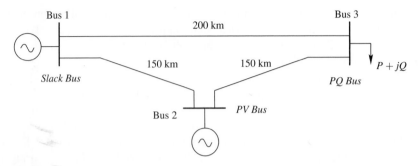

FIGURE 5.1   A three-bus 345-kV example system.

TABLE 5.1   Per-Unit Values in the Example System

Line	Series Impedance Z in Ω (pu)	Total Susceptance B in $\mu\mho$ (pu)
1-2	$Z_{12} = (5.55 + j56.4)\ \Omega = (0.0047 + j0.0474)$ pu	$B_{Total} = 675\mu\mho = (0.8034)$ pu
1-3	$Z_{13} = (7.40 + j75.2)\ \Omega = (0.0062 + j0.0632)$ pu	$B_{Total} = 900\mu\mho = (1.0712)$ pu
2-3	$Z_{23} = (5.55 + j56.4)\ \Omega = (0.0047 + j0.0474)$ pu	$B_{Total} = 675\mu\mho = (0.8034)$ pu

$$Z_{base} = \frac{kV_{base}^2(phase)}{MVA_{base}(1-\phi)} = \frac{kV_{base}^2(L-L)}{MVA_{base}(3-\phi)} = 1190.25\ \Omega \tag{5.1}$$

The base admittance $Y_{base} = 1/Z_{base}$. Based on these base impedance and base admittance, and the line lengths, the line parameters and their per-unit values are as given in Table 5.1 below.

5.4   BUILDING THE ADMITTANCE MATRIX

It is easiest to carry out the network solution in a nodal form, in comparison to writing loop equations, which are more numerous. In writing the nodal equations, the current $\bar{I}_k$ is the current *injected* into a bus k, as shown in Figure 5.2 for the example system of Figure 5.1 [1-3].

By Kirchhoff's Current Law, the current injection at bus k is related to the bus voltages as follows:

$$\bar{I}_k = \bar{V}_k Y_{kG} + \sum_{\substack{m \\ m \neq k}} \frac{\bar{V}_k - \bar{V}_m}{Z_{km}} \tag{5.2}$$

where Y_{kG} is the sum of the admittances connected at bus k to ground, and the second term consists of flows on all the lines connected to bus k, with the series impedance being Z_{km}, for example between buses k and m. Therefore,

$$\bar{I}_k = \bar{V}_k \left(Y_{kG} + \sum_{\substack{m \\ m \neq k}} \frac{1}{Z_{km}} \right) - \sum_{\substack{m \\ m \neq k}} \frac{\bar{V}_m}{Z_{km}} \tag{5.3}$$

In Equation 5.3, the quantities within the brackets on the right side are designated as

$$Y_{kk} = Y_{kG} + \sum_{\substack{m \\ m \neq k}} \frac{1}{Z_{km}} \tag{5.4}$$

which is the self-admittance and is the sum of the admittances connected between bus-k and the other buses, including ground. Similarly in Equation 5.3, between buses k and m, the mutual admittance is

$$Y_{km} = -\frac{1}{Z_{km}} \tag{5.5}$$

which is negative of the inverse of the series impedance between buses k and m. This procedure allows the formulation of the bus-admittance matrix $[Y]$ for an n-bus system as

$$
\begin{bmatrix} \overline{I}_1 \\ \overline{I}_2 \\ .. \\ .. \\ .. \\ \overline{I}_n \end{bmatrix}
=
\begin{bmatrix}
Y_{11} & Y_{12} & .. & .. & .. & Y_{1n} \\
Y_{21} & Y_{22} & .. & .. & .. & Y_{2n} \\
.. & .. & .. & .. & .. & .. \\
.. & .. & .. & .. & .. & .. \\
.. & .. & .. & .. & .. & .. \\
Y_{n1} & Y_{n2} & .. & .. & .. & Y_{nn}
\end{bmatrix}
\begin{bmatrix} \overline{V}_1 \\ \overline{V}_2 \\ .. \\ .. \\ .. \\ \overline{V}_n \end{bmatrix}
\tag{5.6}
$$

Example 5.1
In the example system of Figure 5.2, ignore all the shunt susceptances and assemble the bus-admittance matrix of the form in Equation 5.6.

Solution   In the example system, in per unit, from Table 5.1,

$Z_{12} = (0.0047 + j0.0474)$ pu, $Z_{13} = (0.0062 + j0.0632)$ pu, $Z_{23} = (0.0047 + j0.0474)$ pu

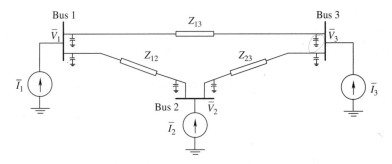

FIGURE 5.2   Example system of Figure 5.1 for assembling Y-bus matrix.

Therefore,

$$
\begin{bmatrix} \bar{I}_1 \\ \bar{I}_2 \\ \bar{I}_3 \end{bmatrix} = \begin{bmatrix} 3.6090 - j36.5636 & -2.0715 + j20.8916 & -1.5374 + j15.6720 \\ -2.0715 + j20.8916 & 4.1431 - j41.7833 & -2.0715 + j20.8916 \\ -1.5374 + j15.6720 & -2.0715 + j20.8916 & 3.6090 - j36.5636 \end{bmatrix} \begin{bmatrix} \bar{V}_1 \\ \bar{V}_2 \\ \bar{V}_3 \end{bmatrix}
$$

5.5   BASIC POWER FLOW EQUATIONS

In the network equation of Equation 5.6, the current injections into buses are not specified explicitly. Rather, the injected real power P is specified at PV buses, and P and Q injections are specified at PQ buses as

$$
P_k + jQ_k = \bar{V}_k \bar{I}_k^*
\tag{5.7}
$$

From the nodal equation in Equation 5.6,

$$
\bar{I}_k = \sum_{m=1}^{n} Y_{km} \bar{V}_m \quad \text{where} \quad (Y_{km} = G_{km} + jB_{km})
\tag{5.8}
$$

From Equation 5.8, recognizing that in terms of complex variables, if $a = bc$, then $a^* = b^* c^*$,

$$
\bar{I}_k^* = \sum_{m=1}^{n} Y_{km}^* \bar{V}_m^* = \sum_{m=1}^{n} (G_{km} - jB_{km}) \bar{V}_m^*
\tag{5.9}
$$

Substituting Equation 5.9 into Equation 5.7,

$$
P_k + jQ_k = \bar{V}_k \bar{I}_k^* = \sum_{m=1}^{n} \left[(G_{km} - jB_{km}) \left(\bar{V}_k \bar{V}_m^* \right) \right]
\tag{5.10}
$$

In Equation 5.10, $\bar{V}_k \bar{V}_m^*$ can be written in the polar form as

$$
\bar{V}_k \bar{V}_m^* = \left(V_k e^{j\theta_k} \right) \left(V_m e^{-j\theta_m} \right) = V_k V_m e^{j\theta_{km}} = V_k V_m (\cos\theta_{km} + j\sin\theta_{km})
\tag{5.11}
$$

where $\theta_{km} = \theta_k - \theta_m$.

Substituting Equation 5.11 into Equation 5.10 and separating real and imaginary parts (note that $\cos\theta_{kk} = 1$ and $\sin\theta_{kk} = 0$),

$$
P_k = G_{kk} V_k^2 + V_k \sum_{\substack{m=1 \\ m \neq k}}^{n} V_m (G_{km} \cos\theta_{km} + B_{km} \sin\theta_{km})
\tag{5.12}
$$

and

$$
Q_k = -B_{kk} V_k^2 + V_k \sum_{\substack{m=1 \\ m \neq k}}^{n} V_m (G_{km} \sin\theta_{km} - B_{km} \cos\theta_{km})
\tag{5.13}
$$

Calculate   for   every   iteration

where the first terms on the rightside of Equations 5.12 and 5.13 correspond to $m = k$. In an n-bus system, let us specify the types of buses as follows: one slack bus, $n_{PV}PV$-buses and $n_{PQ}PQ$-buses such that

$$\underbrace{1}_{\text{slack-bus}} + n_{PV} + n_{PQ} = n \qquad (5.14)$$

In this system we have the following number of equations:

$(n_{PV} + n_{PQ})$ equations: where P are specified

n_{PQ} equations: where Q are specified

Therefore, there are a total of $(n_{PV} + 2n_{PQ})$ equations of the form similar to Equation 5.12 or 5.13.

Similarly, we have the following number of unknowns:

n_{PQ}: unknown voltage magnitudes

$(n_{PV} + n_{PQ})$: unknown voltage phase angles

Therefore, there are a total of $(n_{PV} + 2n_{PQ})$ unknown variables to solve, and as many equations.

There is no closed-form solution to Equations 5.12 and 5.13 that are nonlinear, thus we will use a trial-and-error approach. Using Taylor's series expansion, we will linearize them and use an iterative procedure called the Newton-Raphson method until the solution is reached. To do this, we can write Equations 5.12 and 5.13 as follows where P_k^{sp} and Q_k^{sp} are the specified values of injected real and reactive powers:

$$P_k^{sp} - P_k(V_1, \ldots, V_n, \theta_1, \ldots, \theta_n) = 0 \quad (k \equiv \text{all buses except the slack-bus}) \qquad (5.15)$$

and

$$Q_k^{sp} - Q_k(V_1, \ldots, V_n, \theta_1, \ldots, \theta_n) = 0 \quad (k \equiv \text{all PQ buses}) \qquad (5.16)$$

5.6 NEWTON-RAPHSON PROCEDURE

For solving equations of the form given by Equations 5.15 and 5.16, the Newton-Raphson procedure has now become the commonly used approach due to its speed and the likelihood of conversion, and hence we will only focus on this procedure, recognizing that there are other procedures as well which are used, such as the Gauss-Seidel method described in Appendix A.

The Newton-Raphson procedure is explained by means of a simple nonlinear equation of the same form as Equations 5.15 and 5.16

$$c - f(x) = 0 \qquad (5.17)$$

where c is a constant, and $f(x)$ is a nonlinear function of a variable x. In order to determine the value of x which satisfies Equation 5.17, we will start with an initial guess

$x^{(0)}$ which is not exactly the solution but close to it. A small adjustment Δx is needed so that $(x^{(0)} + \Delta x)$ comes closer to the actual solution:

$$c - f(x^{(0)} + \Delta x) \simeq 0 \qquad (5.18)$$

Using the Taylor series expansion, the function in Equation 5.18 can be expressed as follows where the terms involving higher orders of Δx are ignored:

$$c - \left[f(x^{(0)}) + \frac{\partial f}{\partial x}\Big|_0 \Delta x \right] = 0 \qquad (5.19)$$

or

$$c - f(x^{(0)}) = \frac{\partial f}{\partial x}\Big|_0 \Delta x \qquad (5.20)$$

where the partial derivative is taken at $x = x^{(0)}$. The objective in Equation 5.20 is to calculate the adjustment Δx, which can be calculated as

$$\Delta x = \frac{c - f(x^{(0)})}{\frac{\partial f}{\partial x}\Big|_0} \qquad (5.21)$$

This results in the new estimate of x,

$$x^{(1)} = x^{(0)} + \Delta x \qquad (5.22)$$

Now, similar to Equation 5.21, the new correction to the estimate can be calculated as

$$\Delta x = \frac{c - f(x^{(1)})}{\frac{\partial f}{\partial x}\Big|_1} \qquad (5.23)$$

where the partial derivative is taken at $x = x^{(1)}$. Therefore,

$$x^{(2)} = x^{(1)} + \Delta x \qquad (5.24)$$

This process is repeated until the error $\varepsilon = |c - f(x)|$ is below a certain prespecified error tolerance value and the convergence is assumed to have been reached such that the original equation, Equation 5.17, is satisfied. This Newton-Raphson procedure is illustrated by means of a simple example.

Example 5.2
Consider a simple equation:

$$4 - x^2 = 0 \qquad (5.25)$$

where corresponding to Equation 5.17, $c = 4$, and $f(x) = x^2$. Solve for x which satisfies Equation 5.25 with an error ε below a tolerance of 0.0002.

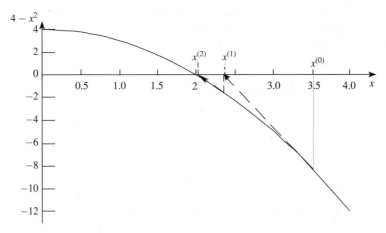

FIGURE 5.3 Plot of $4 - x^2$ as a function of x.

Solution The solution to this equation is obvious: $x = 2$. However, to illustrate the Newton-Raphson procedure, the left side of Equation 5.25 is plotted in Figure 5.3, as a function of x.

For $f(x) = x^2$, $\frac{\partial f}{\partial x} = 2x$. Using this partial derivative and the initial guess as $x^{(0)} = 3.5$, from Equation 5.21,

$$\Delta x = \frac{c - f(x^{(0)})}{\left.\dfrac{\partial f}{\partial x}\right|_0} = -1.17857, \quad x^{(1)} = x^{(0)} + \Delta x = 2.3214 \quad \text{and} \quad \varepsilon = \left|4 - \left(x^{(1)}\right)^2\right| = 0.3214,$$

Using $x^{(1)} = 2.3214$, from Equation 5.23,

$$\Delta x = \frac{c - f(x^{(1)})}{\left.\dfrac{\partial f}{\partial x}\right|_1} = -0.29915, \quad x^{(2)} = x^{(1)} + \Delta x = 2.022 \quad \text{and} \quad \varepsilon = \left|4 - \left(x^{(2)}\right)^2\right| = 0.022$$

Repeating this procedure,

$$\Delta x = \frac{c - f(x^{(2)})}{\left.\dfrac{\partial f}{\partial x}\right|_2} = -0.02188, \quad x^{(3)} = x^{(2)} + \Delta x = 2.000122, \quad \varepsilon = \left|4 - \left(x^{(3)}\right)^2\right| = 0.000122$$

The error $\varepsilon = 0.000122$ is now below the prespecified tolerance and the conversion is achieved.

5.7 SOLUTION OF POWER FLOW EQUATIONS USING N-R METHOD

Having looked at the basics of the Newton-Raphson procedure, we are ready to apply it to the power flow equations of 5.15 and 5.16. In Equations 5.12 and 5.13, P_k and Q_k are in terms of the estimates of voltage magnitudes and phase angles, some of which are unknown, and are thus yet to be determined. In a manner similar to Equation 5.20, we can write the following matrix equation, where corrections are expressed by ΔV and $\Delta \theta$:

$$\underbrace{\begin{bmatrix} P^{sp} - P \\ Q^{sp} - Q \end{bmatrix}}_{(2n_{PQ}+n_{PV}) \times 1} = \underbrace{\begin{bmatrix} \dfrac{\partial P}{\partial \theta} & \dfrac{\partial P}{\partial V} \\ \dfrac{\partial Q}{\partial \theta} & \dfrac{\partial Q}{\partial V} \end{bmatrix}}_{\substack{[J] \\ (2n_{PQ}+n_{PV}) \times (2n_{PQ}+n_{PV})}} \underbrace{\begin{bmatrix} \Delta \theta \\ \Delta V \end{bmatrix}}_{(2n_{PQ}+n_{PV}) \times 1} \tag{5.26}$$

The elements of the $[J]$ matrix above, $\frac{\partial P}{\partial \theta}$ and so on, are submatrices, and $\Delta \theta$ and ΔV are vectors.

In order to evaluate the partial derivatives, we should recognize the following, noting that

$$\theta_{km} = \theta_k - \theta_m, \quad \frac{\partial (\cos \theta_{km})}{\partial \theta_k} = -\sin \theta_{km}, \quad \frac{\partial (\cos \theta_{km})}{\partial \theta_m} = \sin \theta_{km}, \quad \frac{\partial (\sin \theta_{km})}{\partial \theta_k}$$

$$= \cos \theta_{km} \quad \text{and} \quad \frac{\partial (\sin \theta_{km})}{\partial \theta_m} = -\cos \theta_{km}.$$

The steps are as follows:

Step 1: Calculations of $\frac{\partial P}{\partial \theta}$ at all *PV* and *PQ* buses

At a bus k, from Equation 5.12, the partial derivatives with respect to (abbreviated as wrt) θ for the real power are as follows:

$$\frac{\partial P_k}{\partial \theta_k} = V_k \sum_{\substack{m=1 \\ m \neq k}}^{n} V_m(-G_{km} \sin \theta_{km} + B_{km} \cos \theta_{km}), \quad (\text{wrt } \theta_k) \tag{5.27}$$

and

$$\frac{\partial P_k}{\partial \theta_j} = V_k V_j (G_{kj} \sin \theta_{kj} - B_{kj} \cos \theta_{kj}) \quad (\text{wrt } \theta_j, \text{ where } j \neq k) \tag{5.28}$$

Step 2: Calculations of $\frac{\partial P}{\partial V}$ at all *PQ* buses

At a bus k, from Equation 5.12, the partial derivatives with respect to V for the real power are as follows:

$$\frac{\partial P_k}{\partial V_k} = 2G_{kk}V_k + \sum_{\substack{m=1 \\ m \neq k}}^{n} V_m(G_{km} \cos \theta_{km} + B_{km} \sin \theta_{km}) \quad (\text{wrt } V_k) \tag{5.29}$$

and

$$\frac{\partial P_k}{\partial V_j} = V_k (G_{kj} \cos \theta_{kj} + B_{kj} \sin \theta_{kj}) \quad (\text{wrt } V_j \text{ where } j \neq k) \tag{5.30}$$

Step 3: Calculations of $\frac{\partial Q}{\partial \theta}$ at all *PV* and *PQ* buses

At a bus k, from Equation 5.13, partial derivatives with respect to θ for the reactive power are as follows:

$$\frac{\partial Q_k}{\partial \theta_k} = V_k \sum_{\substack{m=1 \\ m \neq k}}^{n} V_m \left(G_{km} \cos \theta_{km} + B_{km} \sin \theta_{km} \right) \quad (\text{wrt } \theta_k) \tag{5.31}$$

and

$$\frac{\partial Q_k}{\partial \theta_j} = V_k V_j \left(-G_{kj} \cos \theta_{kj} - B_{kj} \sin \theta_{kj} \right) \quad (\text{wrt } \theta_j, \text{where } j \neq k) \tag{5.32}$$

Step 4: Calculations of $\frac{\partial Q}{\partial V}$ at all *PQ* buses

At a bus k, from Equation 5.13, partial derivatives with respect to V for the reactive power are as follows:

$$\frac{\partial Q_k}{\partial V_k} = -2B_{kk}V_k + \sum_{\substack{m=1 \\ m \neq k}}^{n} V_m \left(G_{km} \sin \theta_{km} - B_{km} \cos \theta_{km} \right) \quad (\text{wrt } V_k) \tag{5.33}$$

and

$$\frac{\partial Q_k}{\partial V_j} = V_k \left(G_{kj} \sin \theta_{kj} - B_{kj} \cos \theta_{kj} \right) \quad (\text{wrt } V_j, \text{where } j \neq k) \tag{5.34}$$

It should be noted that another row in Equation 5.26 is added for each *PV* bus that becomes a *PQ* bus due to its reactive power reaching one of its limits, if they were to be specified. This is further explained in section 5.10.

Convergence to the Correct Solution: The iterative N-R procedure is continued until all the mismatches in the left-side vector in Equation 5.26 are less than the specified tolerance values, at which point it is assumed that the solution has converged to the correct values. Until the convergence is achieved, in each iterative step, the above matrix equation is solved for the needed corrections as is $\Delta\theta$s and ΔVs. In a practical power network with thousands of buses, the Jacobian J is calculated using sparsity and optimal ordering techniques. This discussion is beyond the scope of this book, and for the example system at hand, we will use the matrix inversion in Equation 5.26 to calculate the correction terms in each iteration step as follows:

$$\begin{bmatrix} \Delta\theta \\ \Delta V \end{bmatrix} = [J]^{-1} \begin{bmatrix} P^{sp} - P \\ Q^{sp} - Q \end{bmatrix} \tag{5.35}$$

Example 5.3

In the example system of Figure 5.1, ignore all the shunt susceptances. Bus-1 is a slack bus, Bus-2 is a *PV* bus, and bus-3 is a *PQ* bus. Using the N-R procedure described above, assemble the Jacobian matrix for the example power system.

Solution   In this system, there are three buses with $n = 3$, $n_{PV} = 1$, and $n_{PQ} = 1$. There are three specified injected real/reactive powers (P_2^{sp}, P_3^{sp} and Q_3^{sp}) and three unknowns (θ_2, θ_3, and V_3) related to the bus voltages. Therefore, the correction terms in Equation 5.34 are of the following form:

$$
\begin{bmatrix} P_2^{sp} - P_2 \\ P_3^{sp} - P_3 \\ Q_3^{sp} - Q_3 \end{bmatrix} = \underbrace{\begin{bmatrix} \dfrac{\partial P_2}{\partial \theta_2} & \dfrac{\partial P_2}{\partial \theta_3} & \dfrac{\partial P_2}{\partial V_3} \\[2mm] \dfrac{\partial P_3}{\partial \theta_2} & \dfrac{\partial P_3}{\partial \theta_3} & \dfrac{\partial P_3}{\partial V_3} \\[2mm] \dfrac{\partial Q_3}{\partial \theta_2} & \dfrac{\partial Q_3}{\partial \theta_3} & \dfrac{\partial Q_3}{\partial V_3} \end{bmatrix}}_{J} \begin{bmatrix} \Delta\theta_2 \\ \Delta\theta_3 \\ \Delta V_3 \end{bmatrix} \tag{5.36}
$$

To assemble the Jacobian matrix J of Equation 5.36, the bus values of k, j, and m in the equations related to the N-R procedure can be recognized for each element as shown in Table 5.2 below:

Equations for the above Jacobian elements are as follows:

$$J(1,1) = V_2 V_1(-G_{21}\sin\theta_{21} + B_{21}\cos\theta_{21}) + V_2 V_3(-G_{23}\sin\theta_{23} + B_{23}\cos\theta_{23}) \tag{5.37}$$

$$J(1,2) = V_2 V_3(G_{23}\sin\theta_{23} - B_{23}\cos\theta_{23}) \tag{5.38}$$

$$J(1,3) = V_2(G_{23}\cos\theta_{23} + B_{23}\sin\theta_{23}) \tag{5.39}$$

$$J(2,1) = V_3 V_2(G_{32}\sin\theta_{32} - B_{32}\cos\theta_{32}) \tag{5.40}$$

$$J(2,2) = V_3 V_1(-G_{31}\sin\theta_{31} + B_{31}\cos\theta_{31}) + V_3 V_2(-G_{32}\sin\theta_{32} + B_{32}\cos\theta_{32}) \tag{5.41}$$

$$J(2,3) = 2G_{33}V_3 + V_1(G_{31}\cos\theta_{31} + B_{31}\sin\theta_{31}) + V_2(G_{32}\cos\theta_{32} + B_{32}\sin\theta_{32}) \tag{5.42}$$

$$J(3,1) = V_3 V_2(-G_{32}\cos\theta_{32} - B_{32}\sin\theta_{32}) \tag{5.43}$$

$$J(3,2) = V_3 V_1(G_{31}\cos\theta_{31} + B_{31}\sin\theta_{31}) + V_3 V_2(G_{32}\cos\theta_{32} + B_{32}\sin\theta_{32}) \tag{5.44}$$

$$J(3,3) = -2B_{33}V_3 + V_1(G_{31}\sin\theta_{31} - B_{31}\cos\theta_{31}) + V_2(G_{32}\sin\theta_{32} - B_{32}\cos\theta_{32}) \tag{5.45}$$

TABLE 5.2   Buses Related to the Jacobian Matrix of Equation 5.36

Equation 5.27	Equation 5.28	Equation 5.30
$\dfrac{\partial P_2}{\partial \theta_2}$: $k = 2$; $m = 1, 3$	$\dfrac{\partial P_2}{\partial \theta_3}$: $k = 2$; $j = 3$	$\dfrac{\partial P_2}{\partial V_3}$: $k = 2$; $j = 3$
Equation 5.28	Equation 5.27	Equation 5.29
$\dfrac{\partial P_3}{\partial \theta_2}$: $k = 3$; $j = 2$	$\dfrac{\partial P_3}{\partial \theta_3}$: $k = 3$; $m = 1, 2$	$\dfrac{\partial P_3}{\partial V_3}$: $k = 3$; $m = 1, 2$
Equation 5.32	Equation 5.31	Equation 5.33
$\dfrac{\partial Q_3}{\partial \theta_2}$: $k = 3$; $j = 2$	$\dfrac{\partial Q_3}{\partial \theta_3}$: $k = 3$; $m = 1, 2$	$\dfrac{\partial Q_3}{\partial V_3}$: $k = 3$; $m = 1, 2$

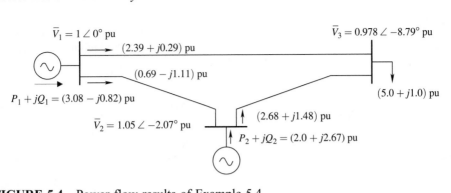

FIGURE 5.4 Power-flow results of Example 5.4.

Example 5.4
In the example system of Figure 5.1, ignore all the shunt susceptances. Bus-1 is a slack bus with $V_1 = 1.0$ pu and $\theta_1 = 0$. Bus-2 is a PV bus with $V_2 = 1.05$ pu and $P_2^{sp} = 2.0$ pu. Bus-3 is a *PQ* bus with injections of $P_3^{sp} = -5.0$ pu and $Q_3^{sp} = -1.0$ pu. Using the N-R procedure described above, the Jacobian matrix was assembled in Example 5.3. Using that in the N-R procedure, calculate the power flow on all three lines in this example power system.

Solution The MATLAB program for this example is included on the accompanying website and the results are as follows:

After the N-R procedure has converged, the Jacobian matrix is as follows:

$$
J = \begin{bmatrix}
43.42 & -21.56 & 0.4237 \\
-21.07 & 35.99 & -1.5913 \\
0.414 & -8.44 & 34.75
\end{bmatrix}
$$

The bus voltages are as follows:

$\overline{V}_1 = 1 \angle 0°$ pu, $\overline{V}_2 = 1.05 \angle -2.07°$ pu, $\overline{V}_3 = 0.978 \angle -8.79°$ pu

At Bus 1, $P_{1-2} + jQ_{1-2} = (0.69 - j1.11)$ pu

At Bus 1, $P_{1-3} + jQ_{1-3} = (2.39 + j0.29)$ pu

At Bus 2, $P_{2-3} + jQ_{2-3} = (2.68 + j1.48)$ pu

The power and reactive powers supplied by the generators at buses 1 and 2 are as follows:

$$P_1 + jQ_1 = (3.08 - j0.82)\text{pu} \quad \text{and} \quad P_2 + jQ_2 = (2.0 + j2.67) \text{ pu}$$

These results are graphically shown in Figure 5.4, which can be verified by a commercial software package such as [4].

5.8 FAST DECOUPLED N-R METHOD FOR POWER FLOW

In power systems, generally reactive powers Qs influence the voltage magnitudes, and the real powers Ps influence the phase angle θs. Therefore, the solution using the

Newton-Raphson method can be considerably simplified by including only the couplings mentioned above, and ignoring the $\partial P/\partial V$ and $\partial Q/\partial\theta$ terms in the Jacobian of Equation 5.26. Thus,

$$[P^{sp} - P] = \left[\frac{\partial P}{\partial\theta}\right][\Delta\theta] \tag{5.46}$$

$$[Q^{sp} - Q] = \left[\frac{\partial Q}{\partial V}\right][\Delta V] \tag{5.47}$$

These two equations are much faster to solve than the coupled set of equations in full Newton-Raphson procedure. Further simplifications can be made such that the elements of the Jacobian matrices of Equations 5.46 and 5.47 are constants and hence do not need to be calculated at every iteration, unlike the scenario in the full N-R method [5]. This technique can be very useful for calculating power flow for contingencies where the speed of calculations is of primary importance, even if the accuracy is somewhat sacrificed compared to the full N-R method.

5.9   SENSITIVITY ANALYSIS

The fast-decoupled formulation of section 5.8 also shows that these equations can be used for sensitivity analysis, for example to determine where to install a reactive power supplying device to control the bus voltage magnitude. This can be seen by rewriting Equation 5.47 as

$$[\Delta Q] = \left[\frac{\partial Q}{\partial V}\right][\Delta V] \tag{5.48}$$

Thus,

$$[\Delta V] = \left[\frac{\partial Q}{\partial V}\right]^{-1}[\Delta Q] \tag{5.49}$$

Equation 5.49 demonstrates the sensitivity of various bus voltage magnitudes to the incremental change in reactive power at a selected bus.

5.10   REACHING THE BUS VAR LIMIT

As discussed in Chapter 9, synchronous generators have limits on the amount of reactive power (var) they can supply. Certain buses may contain additional devices to supply vars but such devices also have their limits. Therefore, in a power-flow condition, if the demand of vars at any *PV* bus reaches its limit, then that bus voltage can be held at its specified magnitude, and thus must be treated as a *PQ* bus.

As an example, in the three-bus system shown in Figure 5.1, bus-2 is a *PV* bus. If the vars needed from the generator at this bus reach the limit, then bus-2 must be treated as a *PQ* bus. Therefore, Equation 5.36 without the limit is modified as follows, where

Q_2^{lim} is the var limit at bus-2 and a column and a row, shown in bold, must be added to the Jacobian matrix:

$$
\begin{bmatrix} P_2^{sp} - P_2 \\ P_3^{sp} - P_3 \\ Q_3^{sp} - Q_3 \\ \boldsymbol{Q_2^{lim} - Q_2} \end{bmatrix} = \underbrace{\begin{bmatrix} \dfrac{\partial P_2}{\partial \theta_2} & \dfrac{\partial P_2}{\partial \theta_3} & \dfrac{\partial P_2}{\partial V_3} & \dfrac{\partial \boldsymbol{P_2}}{\partial \boldsymbol{V_2}} \\[2mm] \dfrac{\partial P_3}{\partial \theta_2} & \dfrac{\partial P_3}{\partial \theta_3} & \dfrac{\partial P_3}{\partial V_3} & \dfrac{\partial \boldsymbol{P_3}}{\partial \boldsymbol{V_2}} \\[2mm] \dfrac{\partial Q_3}{\partial \theta_2} & \dfrac{\partial Q_3}{\partial \theta_3} & \dfrac{\partial Q_3}{\partial V_3} & \dfrac{\partial \boldsymbol{Q_3}}{\partial \boldsymbol{V_2}} \\[2mm] \dfrac{\partial \boldsymbol{Q_2}}{\partial \boldsymbol{\theta_2}} & \dfrac{\partial \boldsymbol{Q_2}}{\partial \boldsymbol{\theta_3}} & \dfrac{\partial \boldsymbol{Q_2}}{\partial \boldsymbol{V_3}} & \dfrac{\partial \boldsymbol{Q_2}}{\partial \boldsymbol{V_2}} \end{bmatrix}}_{J} \begin{bmatrix} \Delta\theta_2 \\ \Delta\theta_3 \\ \Delta V_3 \\ \boldsymbol{\Delta V_2} \end{bmatrix}
\tag{5.50}
$$

In an *n*-bus case, similar modifications are needed at all *PV* buses that reach their var limits and must be treated as *PQ* buses.

5.11 SYNCHRONIZED PHASOR MEASUREMENTS, PHASOR MEASUREMENT UNITS (PMUS), AND WIDE-AREA MEASUREMENT SYSTEMS

In digital relays, being used increasingly for the protection of power systems described in Chapter 13, it is possible to measure the phase angles of the bus voltages at the same instant. These synchronized phasor measurements, in phasor measurement units (PMUs) deployed over a large part of the power systems, known as a wide-area measurement systems, can be used for control, monitoring and protection. These are all part of the smart-grid initiative.

REFERENCES

1. W. D. Stevenson, *Elements of Power System Analysis*, 4th edition, McGraw-Hill, 1982.

2. Prabha Kundur, *Power System Stability and Control*, McGraw-Hill, 1994.

3. Glenn Stagg and A. H. El-Abiad, *Computer Methods in Power System Analysis*, McGraw-Hill, 1968.

4. PowerWorld Computer Program (www.powerworld.com).

5. B. Stott and D. Alsac, "Fast Decoupled Load Flow," IEEE Trans., Vol. PAS-93, 859–869, May/ June 1974.

PROBLEMS

5.1   In Example 5.1, include the line susceptances and construct the bus-admittance *Y* matrix.

5.2   Include the line susceptances in Example 5.4 and compare the results with the solution ignoring them.

5.3   In Example 5.4, ignore the $\partial P/\partial V$ and $\partial Q/\partial \theta$ terms in the Jacobian matrix for a decoupled N-R solution, and compare results and the iterations needed with the full N-R procedure.

5.4   In Example 5.4, calculate the sensitivity of reactive power to voltage at bus-3. Compare the result of using this sensitivity analysis, to injecting 1.0 pu reactive power at bus-3 in the N-R solution of Example 5.4 and computing the increase in bus-3 voltage.

5.5   In Example 5.4, demonstrate the effect of reducing reactive power demand at bus-3 to zero on the bus-3 voltage.

5.6   In Example 5.4, demonstrate the effect of series compensation in line 1-2 in the example three-bus system of Figure 5.1 on the line power flows and bus voltages, where the series reactance of the line is reduced by 50 percent by inserting a capacitor in series with the line 1-2.

5.7   Compute the power flow in Example 5.4 using the Gauss-Seidel method described in the appendix.

POWERWORLD-BASED PROBLEMS

5.8   Calculate the power flow in Example 5.4

5.9   In Problem 5.8, include the line susceptances and compare the results with those of Problem 5.2.

5.10   Compare the results of series compensation with that of Problem 5.6.

5.11   In the system of Example 5.4, the capability to supply vars at bus-2 is limited to 2 pu. Recalculate the power flow.

APPENDIX 5A GAUSS-SEIDEL PROCEDURE FOR POWER FLOW CALCULATIONS

At a PQ-bus k, from Equation 5.6,

$$\overline{I}_k = \sum_{m=1}^{n} Y_{km}\overline{V}_m \tag{A5.1}$$

From Equation 5.7,

$$\overline{I}_k = \frac{P_k - jQ_k}{\overline{V}_k^*} \tag{A5.2}$$

Substituting Equation A5.2 into Equation A5.1,

$$\frac{P_k - jQ_k}{\overline{V}_k^*} = \sum_{m=1}^{n} Y_{km}\overline{V}_m \tag{A5.3}$$

Therefore, rearranging Equation A5.3,

$$\overline{V}_k = \frac{1}{Y_{kk}}\left[\frac{P_k - jQ_k}{\overline{V}_k^*} - \sum_{\substack{m=1 \\ m \neq k}}^{n} Y_{km}\overline{V}_m\right] \tag{A5.4}$$

In the right-hand side of Equation A5.4, P_k and Q_k are specified and the voltage values are the original estimates to calculate the new value of $\overline{V}_k$. New estimated values are used as soon as they are available.

At a *PV*-bus k, where P_k and the voltage magnitude V_k are specified, from Equation A5.3,

$$P_k - jQ_k = \overline{V}_k^* \sum_{m=1}^{n} Y_{km} \overline{V}_m \tag{A5.5}$$

Therefore, using the latest voltage estimates,

$$Q_k = -\mathrm{Im}\left[\overline{V}_k^* \sum_{m=1}^{n} Y_{km} \overline{V}_m \right] \tag{A5.6}$$

The value of Q_k calculated from the equation above is used in Equation A5.4 to get the new estimate of the phase-angle of $\overline{V}_k$ which, keeping the specified voltage magnitude V_k at this *PV*-bus, gives us the new estimate of $\overline{V}_k$. If the value of Q_k calculated by Equation A5.6 is outside the range of the minimum and the maximum reactive power that can be supplied at this bus, then the limit being violated is used in Equation A5.4.

This procedure is repeated until the solution converges.

6

TRANSFORMERS IN POWER SYSTEMS

6.1 INTRODUCTION

Transformers are absolutely essential in making large-scale power transfer feasible over long distances. The primary role of transformers is to change the voltage level. For example, generation in power systems, primarily by synchronous generators, takes place at around 20-kV level. However, it is too low a voltage to transmit economically significant amount of power over long distances. Therefore, transmission voltages of 230 kV, 345 kV, and 500 kV are common and some are as high as 765 kV. At the load-end, these voltages are stepped down to manageable and safe levels such as 120/240 V single-phase in residential usage. Another reason for using transformers in many applications is to provide electrical isolation for safety purposes. Transformers are also needed for converters used in high-voltage DC transmission systems, as discussed in Chapter 7.

Typically in power systems, voltages get transformed approximately five times between generation and the delivery to the ultimate users. Hence, the total installed MVA ratings of transformers are as much as five times larger than that of generators.

6.2 BASIC PRINCIPLES OF TRANSFORMER OPERATION

Transformers consist of two or more tightly coupled windings where almost all of the flux produced by one winding links the other windings. To understand the operating principles of transformers, consider a single coil, also called a winding, of N_1 turns, as shown in Figure 6.1a.

Initially, we will assume that the resistance and the leakage inductance of this winding are both zero; assuming zero leakage flux implies that all of the flux produced by this winding is confined to the core. Applying a time-varying voltage e_1 to this winding results in a flux $\phi_m(t)$. From Faraday's Law:

$$e_1(t) = N_1 \frac{d\phi_m}{dt} \tag{6.1}$$

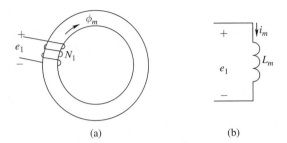

(a)   (b)

FIGURE 6.1 Principle of transformers, beginning with just one coil.

where $\phi_m(t)$ is completely dictated by the time-integral of the applied voltage, as given below (where it is assumed that the flux in the winding is initially zero):

$$\phi_m(t) = \frac{1}{N_1} \int_0^t e_1(\tau) \cdot d\tau \tag{6.2}$$

6.2.1 Transformer Exciting Current

Transformers use Ferromagnetic materials that guide magnetic flux lines and, due to their high permeability, require small ampere-turns (a small current for a given number of turns) to produce the desired flux density. These materials exhibit the multivalued non-linear behavior shown by their *B-H* characteristics in Figure 6.2a.

Imagine that the toroid in Figure 6.1a consists of a ferromagnetic material such as silicon steel. If the current through the coil is slowly varied in a sinusoidal manner with time, the corresponding *H*-field will cause one of the hysteresis loops shown in Figure 6.2a to be traced. Completing the loop once results in a net dissipation of energy within the material, causing power loss referred as the hysteresis loss. Increasing the peak value of the sinusoidally varying *H*-field will result in a bigger hysteresis loop. Joining the peaks of the hysteresis loops, we can approximate the *B-H* characteristic by the single curve shown in Figure 6.2b. In Figure 6.2b, the linear relationship (with a constant μ_m) is approximately valid until the "knee" of the curve is reached, beyond which the material

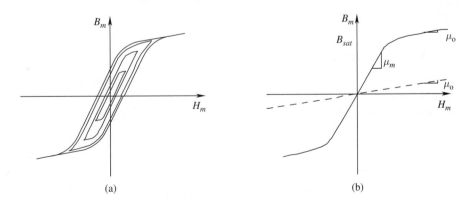

(a)   (b)

FIGURE 6.2 B-H characteristics of ferromagnetic materials.

begins to saturate. Ferromagnetic materials are often operated up to a maximum flux density, slightly above the "knee" of 1.6 T to 1.8 T, beyond which many more ampere-turns are required, yet the flux-density increases only slightly. In the saturated region, the incremental permeability of the magnetic material approaches μ_o, as shown by the slope of the curve in Figure 6.2b.

In the *B-H* curve of the magnetic material in Figure 6.2b, in accordance with Faraday's Law, B_m is proportional to the flux-linkage $\lambda_m (= N_1 \phi_m)$ of the coil, and in accordance with Ampere's Law, H_m is proportional to the magnetizing current i_m drawn by the coil to establish this flux. Therefore, a plot similar to the *B-H* plot in Figure 6.2b can be drawn in terms of λ_m and i_m. In the linear region with a constant slope μ_m, this linear relationship between λ_m and i_m can be expressed by a magnetizing inductance L_m

$$L_m = \frac{\lambda_m}{i_m} \tag{6.3}$$

The current $i_m(t)$ drawn to establish this flux depends on the magnetizing inductance L_m of the winding, as depicted in Figure 6.1b. With a sinusoidal voltage applied to the winding as $e_1 = \sqrt{2}E_1(\text{rms})\cos \omega t$, the core flux from Equation 6.2 is $\phi_m = \hat{\phi}_m \sin \omega t$. In the phasor domain, the relationship between these two, in accordance with the Faraday's Law can be expressed as

$$\sqrt{2}E_1(\text{rms}) = (2\pi f)N_1 \hat{\phi}_m \tag{6.4}$$

or

$$\hat{\phi}_m \simeq \frac{E_1(\text{rms})}{4.44N_1 f} \tag{6.5}$$

which clearly shows that exceeding the applied voltage $E_1(\text{rms})$ above its rated value will cause the maximum flux $\hat{\phi}_m$ to enter the saturation region in Figure 6.2b, resulting in excessive magnetizing current to be drawn. In the saturation region, the magnetizing current drawn is also distorted, with a significant amount of the third-harmonic component.

In modern power transformers of large kVA ratings, the magnetizing currents in their normal operating region are very small—for example, well below 0.2 percent of the rated current at the rated voltage.

6.2.2   Voltage Transformation

A second winding of N_2 turns is now placed on the core, as shown in Figure 6.3a. A voltage is induced in the second winding due to the time-varying flux $\phi_m(t)$ linking it.

From Faraday's Law,

$$e_2(t) = N_2 \frac{d\phi_m}{dt} \tag{6.6}$$

Equations 6.1 and 6.6 show that in each winding, the volts-per-turn are the same due to the same $d\phi_m/dt$:

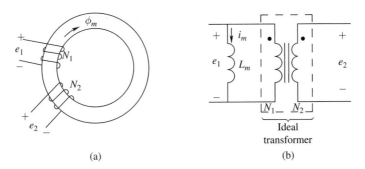

(a) (b)

FIGURE 6.3 Transformer with the open-circuited second coil.

$$\frac{e_1(t)}{N_1} = \frac{e_2(t)}{N_2} \tag{6.7}$$

We can represent the relationship of Equation 6.7 in Figure 6.3b by means of a hypothetical circuit component called the "ideal transformer," which relates the voltages in the two windings by the turns-ratio N_1/N_2 on an instantaneous basis or in terms of phasors in sinusoidal steady state (shown in brackets):

$$\frac{e_1(t)}{e_2(t)} = \frac{N_1}{N_2} \quad \text{and} \quad \left(\frac{\overline{E}_1}{\overline{E}_2} = \frac{N_1}{N_2} \right) \tag{6.8}$$

The dots in Figure 6.3b convey the information that the winding voltages will be of the same polarity at the dotted terminals with respect to their undotted terminals. For example, if ϕ_m is increasing with time, the voltages at both dotted terminals will be positive with respect to the corresponding undotted terminals. The advantage of using this dot convention is that the winding orientations on the core need not be shown in detail.

A load such an *R-L* combination is now connected across the secondary winding, as shown in Figure 6.4a. A current $i_2(t)$ will now flow through the load. The resulting ampere-turns $N_2 i_2$ will tend to change the core flux ϕ_m but *cannot* because $\phi_m(t)$ is completely dictated by the time-integral of the applied voltage $e_1(t)$, as given in Equation 6.2. Therefore,

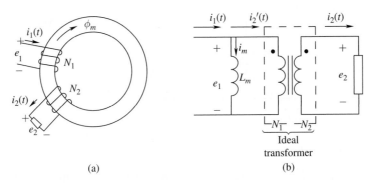

(a) (b)

FIGURE 6.4 Transformer with load connected to the secondary winding.

additional current i_2' in Figure 6.4b is drawn by winding 1 in order to compensate (or nullify) N_2i_2, such that $N_1i_2' = N_2i_2$:

$$\frac{i_2'(t)}{i_2(t)} = \frac{N_2}{N_1} \quad \text{and} \quad \left(\frac{\bar{I}_2'}{\bar{I}_2} = \frac{N_2}{N_1}\right) \tag{6.9}$$

This is the second property of the "ideal transformer." Thus, the total current drawn from the terminals of winding 1 is

$$i_1(t) = i_m(t) + i_2'(t) \quad \text{and} \quad \left(\bar{I}_1 = \bar{I}_m + \bar{I}_2'\right) \tag{6.10}$$

6.2.3   Transformer Equivalent Circuit

In Figure 6.4b, the resistance and the leakage inductance associated with winding 2 appear in series with the *R-L* load. Therefore, the induced voltage e_2 differs from the voltage v_2 at the winding terminals by the voltage drop across the winding resistance and the leakage reactance as depicted in Figure 6.5 in the phasor domain. Similarly, the applied voltage v_1 differs from the emf e_1 (induced by the time-rate of change of the flux ϕ_m) in Figure 6.4b by the voltage drop across the resistance and the leakage inductance of winding 1, represented in Figure 6.5 in the phasor-domain.

6.2.4   Core Losses

The loss due to the hysteresis-loop in the *B-H* characteristic of the magnetic material was discussed earlier. Another source of core loss is due to eddy currents. All magnetic materials have a finite electrical resistivity (ideally, it should be infinite). By Faraday's voltage induction law, time-varying fluxes induce voltages in the core, which result in circulating (eddy) currents within the core to oppose these flux changes (and partially neutralize them).

In Figure 6.6a, an increasing flux ϕ will set up many current loops (due to induced voltages that oppose the change in core flux), which result in losses. The primary means of limiting the eddy-current losses is to make the core out of steel laminations which are insulated from each other by means of thin layers of varnish, as shown in Figure 6.6b.

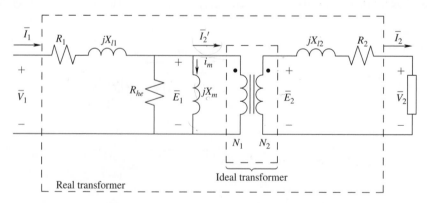

FIGURE 6.5   Transformer equivalent circuit including leakage impedances and core losses.

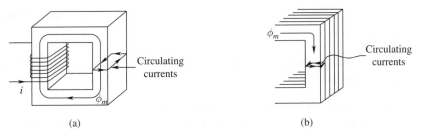

(a) (b)

FIGURE 6.6 Eddy currents in the transformer core.

A few laminations are shown to illustrate how insulated laminations reduce eddy-current losses. Because of the insulation between laminations, the current is forced to flow in much smaller loops within each lamination. Laminating the core reduces the flux and the induced voltage more than it reduces the effective resistance to the currents within a lamination, thus reducing the overall losses. For 50- or 60-Hz operation, lamination thicknesses are typically 0.2 to 1 mm. We can model core losses due to hysteresis and eddy currents by connecting a resistance R_{he} in parallel with X_m, as shown in Figure 6.5. In large modern transformers, core losses are well below 0.1 percent of the transformer MVA rating.

6.2.5 Equivalent Circuit Parameters

In order to utilize the transformer equivalent circuit of Figure 6.5, we need the values of its various parameters. These specifications are generally provided by the manufacturers of power transformers. These can also be obtained using short-circuit and open-circuit tests. In the open-circuit test, the rated voltage is applied to the low-voltage winding, keeping the high-side open-circuited. This allows the magnetizing reactance and the core-loss equivalent resistance to be estimated. In the short-circuit test, the low-voltage winding is short-circuited and a reduced voltage is applied to the high-voltage winding that results in the rated current. This allows the leakage impedances in the transformer equivalent circuit to be estimated. A detailed discussion of the open-circuit and the short-circuit tests can be found in any of the basic books dealing with transformers.

6.3 SIMPLIFIED TRANSFORMER MODEL

Consider the equivalent circuit of a real transformer, shown in Figure 6.5. In many power system studies, the excitation current, which is the sum of the magnetizing current and the core-loss current components, is neglected resulting in the simplified model shown in Figure 6.7. In this model, the subscript p refers to the primary winding and s to the

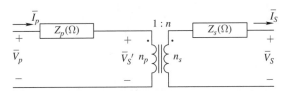

FIGURE 6.7 Simplified transformer model.

secondary winding. Z_p and Z_s are the leakage impedances of the primary and the secondary windings, and the turns-ratio $n = n_s/n_p$.

6.3.1 Transferring Leakage Impedances across the Ideal Transformer Portion

In the simplified model of Figure 6.7, if the secondary-winding terminals are hypothetically short-circuited, then $\overline{V}_s = 0$, and related by the ideal transformer turns ratio,

$$\overline{V}'_s = \frac{Z_s \overline{I}_s}{n} \quad \text{(under short circuit)} \tag{6.11}$$

Also related by the ideal transformer turns ratio, $\overline{I}_s = \overline{I}_p/n$. Substituting for $\overline{I}_s$ in Equation 6.11,

$$\overline{V}'_s = \left(\frac{Z_s}{n^2}\right)\overline{I}_p \quad \text{(under short circuit)} \tag{6.12}$$

In Figure 6.7,

$$\overline{V}_p = \overline{V}'_s + Z_p \overline{I}_p \tag{6.13}$$

and therefore, from the primary-winding terminals under this hypothetical short-circuit, the impedance "seen" from the primary side is $\overline{V}_p/\overline{I}_p$, which by using Equations 6.12 and 6.13, is

$$Z_{ps}(\Omega) = Z_p + \left(Z_s/n^2\right) \tag{6.14a}$$

as shown in Figure 6.8a. Similarly, if the primary-winding leakage impedance is transferred to the secondary winding, then

$$Z_{sp}(\Omega) = Z_s + \left(n^2 Z_p\right) \tag{6.14b}$$

as shown in Figure 6.8b.

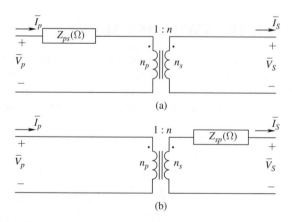

(a)

(b)

FIGURE 6.8 Transferring leakage impedances across the ideal portion of the transformer model.

6.4   PER-UNIT REPRESENTATION

Most power system studies such as power flow, transient analysis, and short-circuit fault calculations are carried out in terms of per-unit. In a transformer, the rated voltages and currents on each side are considered as the base values. Since the voltages and currents on the two sides are related by the turns ratio, the MVA base, which is the product of the voltage and current bases, is the same on each side. In terms of base values, the base impedance magnitudes $Z_{p,base}$ and $Z_{s,base}$ are

$$Z_{p,base} = V_{p,rated}/I_{p,\,rated} \quad \text{and} \quad Z_{s,base} = V_{s,rated}/I_{s,rated} \tag{6.15}$$

The magnitudes of voltage and currents associated with the ideal transformer in Figure 6.7 are related as follows:

$$\frac{V_{p,rated}}{V_{s,rated}} = \frac{1}{n} \quad \text{and} \quad \frac{I_{p,rated}}{I_{s,rated}} = n \tag{6.16}$$

Therefore, the base impedance magnitudes on the two sides, given by Equation 6.15, are related as

$$\frac{Z_{p,base}}{Z_{s,base}} = \left(\frac{1}{n}\right)^2 = \left(\frac{n_p}{n_s}\right)^2 \tag{6.17}$$

Using these base values, all the parameters and the variables in Figures 6.8a and b can be expressed in per unit as the ratio of their base values, as shown by a common equivalent circuit of Figure 6.9, where $Z_{tr}(pu)$ is the transformer per-unit leakage impedance, equal to $Z_{ps}(pu)$ and $Z_{sp}(pu)$:

$$Z_{tr}(\text{pu}) = \underbrace{Z_{ps}(\text{pu})}_{\left(=\frac{Z_{ps}}{Z_{p,base}}\right)} = \underbrace{Z_{sp}(\text{pu})}_{\left(=\frac{Z_{sp}}{Z_{s,base}}\right)} \tag{6.18}$$

In Figure 6.9, the primary and secondary winding currents are equal in per unit—that is, $\bar{I}_p(\text{pu}) = \bar{I}_s(\text{pu}) = \bar{I}(\text{pu})$—and the voltages on the two sides differ by the voltage drop across the leakage impedance.

In three-phase power systems, transformer windings are connected in a wye or a delta arrangement as shown in Figure 6.10. The example below shows their per-phase

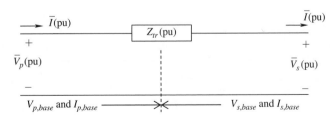

FIGURE 6.9   Transformer equivalent circuit in per unit (pu).

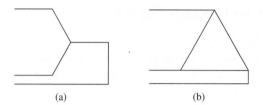

(a)   (b)

FIGURE 6.10   Winding connections in a three-phase system.

representation in per-unit assuming that the primary-side and the secondary-side windings are connected as wye-wye or delta-delta. If one side is wye-connected and the other-side delta-connected, then there is a phase-shift of 30° which must be taken into account, as discussed later on.

Example 6.1

Consider that the 200-km long transmission line between buses 1 and 3 in the three-bus power system of Chapter 5 in Figure 5-1 is at 500 kV. Two 345/500 kV transformers are used at both ends, as shown in the one-line diagram of Figure 6.11. In the per-unit study, if the system base voltage is 345 kV, then the line impedance and the transformer leakage impedances are all calculated on the 345-kV voltage base, and 100 MVA base. For this 500-kV transmission line, the series reactance is 0.326 Ω/km at 60 Hz and the series resistance is 0.029 Ω/km. Neglect line susceptances. Each of the transformers has a leakage reactance of 0.2 pu on the 1000-MVA base.

Calculate the per-unit series impedance of the transmission line and the transformers between buses 1 and 3 for the power flow study as discussed in Chapter 5, where the line-line voltage base is 345 kV and the three-phase MVA base is 100 MVA.

Solution   The 500-kV transmission line is 200-km long. From the given parameter values, the series impedance of the line is $Z_{Line} = (5.8 + j65.2)\ \Omega$. In a three-phase system,

$$Z_{base}(\Omega) = \frac{kV_{base}^2(L\text{-}L)}{MVA_{base}(3\text{-phase})} \tag{6.19}$$

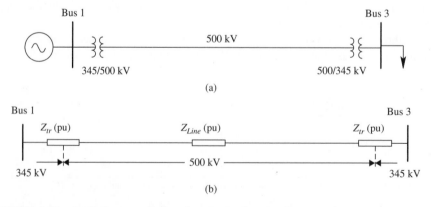

(a)

(b)

FIGURE 6.11   Including nominal-voltage transformers in per-unit.

and therefore, at the 500-kV voltage and 1,000 MVA (three-phase) basis, the base impedance is $Z_{base} = 250.0 \ \Omega$. Therefore, in per-unit, the series impedance of the transmission line is $Z_{Line} = (0.0232 + j0.2608)$ pu. Each of the transformer impedances are given as $Z_{tr} = j0.2$ pu. All the impedances in per-unit, on a 1,000 MVA basis are shown in Figure 6.11b.

Using Equation 6.18, the transmission-line impedance can be represented on the 345-kV side of either transformer and its per-unit value will not change. Therefore, the impedance between buses 1 and 3 on a 345-kV and 1000-MVA base is as follows in the diagram of Figure 6.11b:

$$Z_{13} = j0.2 + (0.0232 + j0.2608) + j0.2 = (0.0232 + j0.6608) \text{ pu}$$

We now need to express this impedance on 345-kV and 100-MVA base for use in power-flow studies of Chapter 5. Making use of Equation 6.19, the per-unit impedance from an original MVA base to a new MVA base is as follows:

$$Z_{pu}(\text{new}) = Z_{pu}(\text{original}) \times \frac{MVA_{base}(\text{new})}{MVA_{base}(\text{original})} \tag{6.20}$$

Therefore, from Equation 6.20, using 100-MVA as the new base and 1000 MVA as the original base, the series impedance between buses 1 and 3 is

$$Z_{13} = (0.00232 + j0.06608) \text{ pu}$$

6.5 TRANSFORMER EFFICIENCIES AND LEAKAGE REACTANCES

Power transformers are designed to minimize power losses within them. These consist of the I^2R losses, also called the copper losses, in the windings, and the core losses. Core losses are mostly independent of the transformer loading, where as the losses in windings depend on the square of the transformer loading. The energy efficiency of a transformer, in fact of any apparatus, is defined as follows, where the output power equals the input power minus the losses:

$$\%\text{Efficiency} = 100 \times \frac{P_{output}}{P_{input}} = 100 \times \left(1 - \frac{P_{losses}}{P_{input}}\right) \tag{6.21}$$

Generally, transformer efficiencies are at their maximum at a load when the core losses and the winding losses are equal to each other. In large power transformers, these efficiencies are generally in excess of 99.5% at or near full-load.

For achieving high efficiencies in transformers, their winding resistances are generally well below 0.5% or 0.005 pu. Therefore, their leakage impedances are dominated by the leakage reactances, which depend of the voltage class of transformers. These leakage reactances are approximately in the following ranges: 7–10 percent in 69-kV transformers, 8–12 percent in 115-kV transformers, and 11–16 percent in 230-kV transformers. In 345-kV and 500-kV transformers, these reactances are 20 percent or even larger.

6.6 REGULATION IN TRANSFORMERS

Large values of leakage reactances are helpful in reducing currents during power system faults like short-circuits on transmission lines to which these transformers are connected. However, for a constant input voltage applied, the output voltage of a transformer changes, based on the transformer loading, due to the voltage drop across its leakage impedance. This is called *Regulation*, which is the change in output voltage, expressed as the percentage of the rated output voltage, if the rated output kVA load, at a specified power factor, is reduced to zero (that is, the secondary winding is open-circuited). From Figure 6.9, assuming the leakage impedance to be purely reactive:

$$\overline{V}_s(\text{pu}) = \overline{V}_P(\text{pu}) - jX_{tr}(\text{pu})\overline{I}(\text{pu}) \tag{6.22}$$

It can be observed from Equation 6.22 that lower the power-factor (lagging) of the load, larger the change in the output voltage magnitude and hence larger the regulation.

6.6.1 Transformers Tap-Changing for Voltage Control

By tap-changing in transformers, it is possible to adjust the output voltage magnitude. It is possible to change taps under load, and such arrangements, called load tap changers (LTC) are described in detail in [1]. Tap-changing is usually accomplished using auto-transformers discussed in the next section. Tap-changing can be included in power flow studies, as illustrated by a homework problem using *PowerWorld*.

6.7 AUTO-TRANSFORMERS

Auto-transformers are very frequently used in power systems for transforming voltages where electrical isolation is not necessary, and for tap-changing. In this analysis, we will assume an ideal transformer, neglecting the leakage impedances and the excitation current. Therefore, all the voltages and the currents are of the same phase respectively, and hence are represented by their magnitudes alone for convenience, rather than as phasors.

In the two-winding arrangement shown in Figure 6.12a, in terms of the rated voltage and the rated current associated with each winding, the transformer rating is

$$\text{Two-Winding Transformer Rating} = V_1 I_1 = V_2 I_2 \tag{6.23}$$

In the autotransformer arrangement shown in Figure 6.12b, these two winding are connected in series. With V_1 applied to the low-side on the left, the high-side voltage is $(V_1 + V_2)$, as shown in Figure 6.12b. The high-side is rated at I_2 without overloading any of the windings, and at this condition, from Kirchhoff's Current Law, the current on the low-side is $(I_1 + I_2)$. The product of the high-side voltage and current ratings is (using the product on the low-side will yield similar results) the auto-transformer rating—that is, the volt-amperes that can be transferred through it:

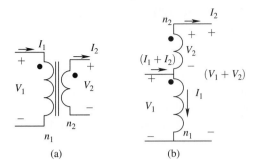

FIGURE 6.12 Auto-transformer.

$$\text{Auto-Transformer Rating} = (V_1 + V_2)I_2 \tag{6.24}$$

Comparing Equations 6.23 and 6.24, the ratings of the two arrangements are related as

$$\text{Auto-Transformer Rating} = \left(1 + \frac{V_1}{V_2}\right) \times \text{Two-Winding Transformer Rating} \tag{6.25}$$

Since the two-winding transformer rating equals the system VA transfer requirement, the needed auto-transformer is equivalent to a two-winding transformer of the following rating:

Equivalent Two-Winding Transformer Rating

$$= \text{Transfer Requirement} \div \left(1 + \frac{V_1}{V_2}\right) \tag{6.26}$$

From Figure 6.12b, expressing the term in the bracket in Equation 6.26 in terms of the high-side voltage $V_H (= V_1 + V_2)$ and the low-side voltage $V_L (= V_1)$:

Equivalent Two-Winding Transformer Rating

$$= \left(1 - \frac{V_L}{V_H}\right) \times \text{Transfer Requirement} \tag{6.27}$$

If the high-side and the low-side voltages are not far apart, the equivalent two-winding transformer rating of an auto-transformer can be much smaller than the transfer requirement.

Example 6.2
In a system, 1 MVA has to be transferred with the low-side voltage of 22 kV and the high-side voltage of 33 kV [2]. Calculate the equivalent two-winding transformer rating of an auto-transformer to satisfy this requirement.

Solution   From Equation 6.27, it is 333 kVA.

The example above shows that only a 333-kVA auto-transformer will suffice, which otherwise would require a 1,000 kVA conventional two-winding transformer. Therefore, an auto-transformer will be physically smaller and less costly. Efficiencies of auto-transformers are higher than their two-winding counterparts, whereas their leakage reactances are comparatively smaller. The main disadvantage of auto-transformers is that there is no electrical (galvanic) isolation between the two sides—however, it is not always needed. Hence, auto-transformers are very often used in power systems.

6.8 PHASE-SHIFT INTRODUCED BY TRANSFORMERS

Primarily, there are two ways that phase shift in voltages are introduced by transformer-connections in three-phase power systems. One common practice is to connect transformers in a wye on one-side, and in a delta on the other-side, which results in a phase-shift of 30° in voltages between the two sides. Another type of transformer-connection is where controllable phase-shift is desired to regulate the power flow through the transmission line to which the transformer is connected. We will look at both of these cases.

6.8.1 Phase-Shift in Δ-Y Transformers

Transformers connected in Δ-Y, as shown in Figure 6.13a, result in a 30° phase-shift that is shown by the phasor diagram of Figure 6.13b. In order to boost the voltages produced by the generators, the low-voltage sides are connected in a delta and the high-voltage sides are connected in a grounded wye. On the Δ-connected side, the terminal voltages, although isolated from ground, can be visualized by hypothetically connecting very large, but equal, resistances from each terminal to a hypothetical neutral n. Thus, $\overline{V}_{An}$ is the terminal-A voltage with respect to the hypothetical neutral n. As shown in Figure 6.13b, $\overline{V}_{An}$ leads $\overline{V}_a$ by 30° and the magnitude of the two voltages can be related as follows:

$$\overline{V}_{AC} = \left(\frac{n_1}{n_2}\right)\overline{V}_a, \text{ and thus } \overline{V}_{An} = \frac{1}{\sqrt{3}}\left(\frac{n_1}{n_2}\right)\overline{V}_a e^{j30°} \qquad (6.28)$$

Based on Equation 6.28, the per-phase equivalent circuit is as shown in Figure 6.13c.

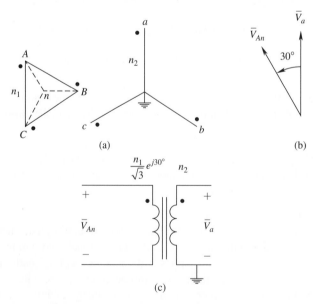

(a)   (b)

(c)

FIGURE 6.13 Phase-shift in Δ-Y connected transformers.

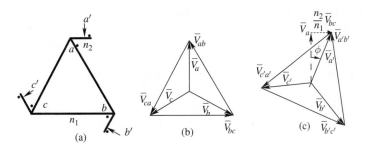

FIGURE 6.14 Transformer for phase-angle control.

6.8.2 Phase-Angle Control

As discussed in Chapter 2, power flow between two AC systems depends on sin δ, where δ is difference of the phase angles between the two. Figure 6.14a shows a phase-regulating transformer arrangement for achieving a phase shift.

Phasor diagram of the incoming voltages $a - b - c$ is shown in Figure 6.14b and of the outgoing voltages $a' - b' - c'$ in Figure 6.14c, which show that the outgoing voltages lag by an angle ϕ with respect to the incoming voltages. The sliders in Figure 6.14a represent finite taps.

6.9 THREE-WINDING TRANSFORMERS

Often in power system transformers and auto-transformers, an isolated third winding is added to which capacitors or power-electronics-based apparatus to supply reactive power to support the system voltage are connected. The Δ-connected tertiary winding is usually of a low voltage rating, and of a much lower MVA rating. The purpose of the tertiary winding is to provide a path for the zero-sequence currents to circulate, as discussed in the chapter on transmission-line faults, and to connect devices to supply reactive power.

The operating principle of the three-winding transformers is an extension of the two-winding transformers discussed earlier. Often, an auto-transformer arrangement is used between the high-side and the low-side, and a tertiary winding is added and connected as shown in Figure 6.15a. These can be represented as shown in Figure 6.15b.

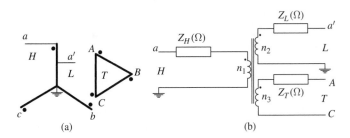

FIGURE 6.15 Three-winding auto-transformer.

6.10   THREE-PHASE TRANSFORMERS

As discussed earlier, power transformers consist of three phases, where separate single-phase transformers discussed earlier may be connected in a Y or a Δ arrangement. In contrast, in a three-phase transformer, all the windings are placed on a common core, thus resulting in reduced core cost. The decision on the type of transformer to purchase depends on such factors as initial installed cost, maintenance costs, operating cost (efficiency), reliability, etc. Three-phase units have lower construction and maintenance costs and can be built to the same efficiency ratings as single-phase units. Many electrical systems have mobile substations and emergency spare transformers to provide backup in case of failure [4].

In extra-high-voltage systems at high power levels, single-phase transformers are used. The reason has to do with the size of the three-phase transformers at these levels makes their transportation from the manufacturing site to the installation site difficult. For unique requirements it may be economical to have emergency spare single-phase transformers (one-third of the total rating) rather than having spare three-phase transformers [4].

Three-phase transformers are either a shell or a core type. In balanced three-phase operation, the equivalent circuit of a three-phase transformer of either type, on a per-phase basis, is similar to that of single-phase transformers. Only under unbalanced operation, such as during unsymmetrical faults, the difference between these three-phase transformer types appears.

6.11   REPRESENTING TRANSFORMERS WITH OFF-NOMINAL TURNS RATIOS, TAPS, AND PHASE-SHIFT

As shown by Example 6.1, if a transformer with nominal turns ratio and without phase-shift is encountered, then it can be simply represented by its leakage impedance. For transformers with off-nominal turns-ratio, taps, and phase-shift, special consideration must be given. Figure 6.16 shows a general representation where the voltage transformation is by $1:t$, where t is a real number in case of off-nominal turns ratio and a complex number in case of a phase-shifter; Z_ℓ is the leakage impedance and $Y_\ell = 1/Z_\ell$.

Considering the leakage impedance separately, the voltage transformation across the ideal-transformer portion is as shown in Figure 6.16 and the sum of the complex powers into it must be equal to zero so that

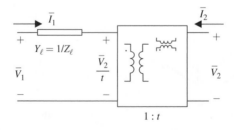

FIGURE 6.16   General representation of an auto-transformer and a phase-shifter.

$$\frac{\overline{V}_2}{t}\overline{I}_1^* = -\overline{V}_2\overline{I}_2^* \tag{6.29}$$

In Figure 6.16,

$$\overline{I}_1 = \left(\overline{V}_1 - \frac{\overline{V}_2}{t}\right)Y_\ell \tag{6.30}$$

From Equations 6.29 and 6.30, and recognizing that $t\,t^* = |t|^2$,

$$\overline{I}_2 = -\frac{\overline{I}_1}{t^*} = -\overline{V}_1\frac{Y_\ell}{t^*} + \overline{V}_2\frac{Y_\ell}{|t|^2} \tag{6.31}$$

From Equations 6.30 and 6.31,

$$\begin{bmatrix} \overline{I}_1 \\ \overline{I}_2 \end{bmatrix} = \begin{bmatrix} Y_\ell & -\dfrac{Y_\ell}{t} \\ -\dfrac{Y_\ell}{t^*} & \dfrac{Y_\ell}{|t|^2} \end{bmatrix} \begin{bmatrix} \overline{V}_1 \\ \overline{V}_2 \end{bmatrix} \tag{6.32}$$

This nodal representation is similar to the one discussed in Chapter 5 for representing the system for power flow studies. It is discussed below, separately in auto-transformers and phase-shifters.

6.11.1 Off-Nominal Turns-Ratios and Taps

The per-unit representation of transformers with off-nominal turns-ratio and taps (without phase shift) is by a transformer with a turns-ratio $1 : t$ where t is real in Figure 6.17a. Equation 6.32 can be represented by a pi-circuit as shown in Figure 6.17b, where $t^* = t$ since t is real. The shunt admittances in Figure 6.17b can be calculated to be as shown.

Example 6.3
In the transformer of Figure 6.18a in per unit, the leakage impedance $Z_\ell = j0.1$ pu. The turns-ratio $t = 1.1$ in per-unit in order to boost the voltage at side-2. Calculate the parameters in the pi-circuit of Figure 6.18b for use in power flow programs.

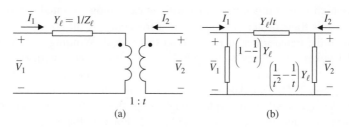

(a) (b)

FIGURE 6.17 Transformer with an off-nominal turns-ratio or taps in per unit; t is real.

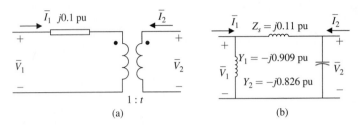

FIGURE 6.18   Transformer of Example 6.3.

Solution   In this example, $Y_\ell = 1/Z_\ell = -j10$ pu. In the pi-circuit, $Y_1 = -j0.909$ pu, where the susceptance value is negative, implying that it is inductive. $Y_2 = j0.826$ pu, where the susceptance value is positive, implying that it is capacitive. The corresponding pi-circuit model is shown in Figure 6.18b, which clearly shows how the LC circuit results in higher V_2 than V_1 in magnitude.

6.11.2   Representing Transformer Phase-Shift

In a phase-shift transformer arrangement, t in the Y-matrix of Equation 6.32 is complex. Therefore, $Y_{12} \neq Y_{21}$. As a consequence, a transformer with a complex turns-ratio cannot be represented by a pi-circuit; rather, it calls for an admittance matrix of Equation 6.32 to be used in the power flow studies.

REFERENCES

1. Prabha Kundur, *Power System Stability and Control*, McGraw Hill, 1994.
2. *Electrical Transmission and Distribution Reference Book*, Westinghouse Electric Corporation, 1950.
3. P. Anderson, *Analysis of Faulted Power Systems*, June 1995, Wiley-IEEE Press.
4. United States Department of Agriculture, Rural Utilities Service, *Design Guide for Rural Substations*, RUS BULLETIN 1724E-300 (http://www.rurdev.usda.gov/RDU_Bulletins_ Electric.html).

PROBLEMS

6.1   A transformer is designed to step down the applied voltage of 2400 V (rms) to 240 V (rms) at 60 Hz. Calculate the maximum rms voltage that can be applied to the high side of this transformer without exceeding the rated flux density in the core if this transformer is supplied by a frequency of 50 Hz.

6.2   Assume the transformer in Figure 6.4a to be ideal. Winding 1 is applied a sinusoidal voltage in steady state with $\overline{V}_1 = 120$ V $\angle 0°$ at a frequency $f = 60$ Hz. $N_1/N_2 = 3$. The load on winding 2 is a series combination of R and L with $Z_L = (5 + j3)$ Ω. Calculate the current drawn from the voltage source.

6.3   Consider an ideal transformer, neglecting the winding resistances, leakage inductances, and the core loss. $N_1/N_2 = 3$. For a voltage of 120 V (rms) at a frequency of 60 Hz applied to winding 1, the magnetizing current is 1.0 A (rms).

If a load of 1.1 Ω at a power factor of 0.866 (lagging) is connected to the secondary winding, calculate $\bar{I}_1$.

6.4 A 2400/240-V, 60-Hz transformer has the following parameters in the equivalent circuit of Figure 6.5: the high-side leakage impedance is $(1.2 + j\,2.0)$ Ω, the low-side leakage impedance is $(0.012 + j\,0.02)$ Ω, and Xm at the high-side is 1800 Ω. Neglect R_{he}. Calculate the input voltage if the output voltage is 240 V (rms) and supplying a load of 1.5 Ω at a power factor of 0.9 (lagging).

6.5 Calculate the equivalent-circuit parameters of a transformer, if the following open-circuit and short-circuit test data is given for a 60-Hz, 50-kVA, 2400:240-V distribution transformer:

open-circuit test with high-side open: $Voc = 240$ V, $Ioc = 5.0$ A, $Poc = 400$ W,

short-circuit test with low-side shorted: $Vsc = 90$ V, $Isc = 20$ A, $Psc = 700$ W.

6.6 Three two-winding transformers are grounded Y-Δ connected to 230 kV on the Y-side and to 34.5 kV on the Δ-side. The combined three-phase rating of these transformers is 200 MVA. The per-unit reactance is 11 percent on the basis of the transformer rating. Calculate the rated values of voltage and currents, and X_{ps} and X_{sp} (in Ω) in Figures 6.8a and b.

6.7 In a transformer, the leakage reactance is 9 percent on the basis of its rating. What will be its percent value if, simultaneously the voltage-base is doubled and the MVA-base is halved.

6.8 Efficiency in a power transformer is 98.6 percent when loaded such that its core losses equal the copper losses. Calculate the per unit resistance of this transformer.

6.9 Using Figure 6.9, calculate $\bar{V}_p$ in per unit, if $\bar{V}_s(\text{pu}) = 1\,\angle 0^0$, rated $\bar{I}(\text{pu}) = 1\,\angle -30°$, and $X_{tr} = 11$ percent. Draw the relationship between these calculated variables by a phasor diagram.

6.10 The % regulation for a transformer is defined as

$$\% \text{ Regulation} = 100 \times \frac{V_{\text{no-load}} - V_{\text{rated}}}{V_{\text{rated}}},$$

where V_{rated} is the voltage at the rated kVA loading at the specified power load. Calculate the % regulation for the transformer in Problem 6.9.

6.11 Calculate the regulation in problem 6.10, if the rated $\bar{I}(\text{pu}) = 1\,\angle 0°$. Compare this value with that in problem 6.10.

6.12 Derive the expression for the full-load regulation, where θ is the power factor angle, which is taken as positive when the current lags the voltage, and X_{tr} is in per-unit.

6.13 In Example 6.2, the auto-transformer is loaded to its rated MVA at unity power factor. Calculate all voltages and currents in the auto-transformer in Figure 6.12b.

6.14 In Example 6.2, if a two-winding transformer is selected, the efficiency of the two-winding transformer is 99.1 percent. If the same transformer is connected as an auto-transformer, calculate its efficiency based on the MVA rating that it is capable of. Assume a unity power factor of operation in both cases.

6.15 In a Y-Δ connected transformer as shown in Figure 6.13a, show all the phase and line-line voltages on both sides of the transformer, similar to Figure 6.13b. The line-line voltage on Y-side is 230 kV and on the Δ-side 34.5 kV. Assume the phase angle $\bar{V}_a$ to 90°.

6.16   In problem 6.15, the wye-connected load on the Δ-side is purely resistive and the load-current amplitude is 1 pu. Draw the currents within the Δ windings and those drawn from the Y-side in per-unit.

TRANSFORMERS IN PSCAD/EMTDC

6.17   Obtain the transformer waveforms, as described on the accompanying website.

INCLUDING TRANSFORMERS IN POWER FLOW STUDIES USING *POWERWORLD*

6.18   The power flow in a three-bus example power system is calculated using MATLAB in Example 5-4. Repeat this using *PowerWorld*, where include an auto-transformer in this example for voltage regulation, as described in the accompanying website, and calculate the results of the power flow study.

6.19   The power flow in a three-bus example power system is calculated using MATLAB in Example 5-4. Repeat this using *PowerWorld*, where include a phase-shifting transformer in this example for controlling power flow, as described in the accompanying website, and calculate the results of the power flow study.

7

HIGH VOLTAGE DC (HVDC) TRANSMISSION SYSTEMS

7.1 INTRODUCTION

For transmitting large amounts of power over long distances, for example from a remote source to a load center far away, a high voltage DC (HVDC) system can be more economical than an AC transmission system. In other words, an HVDC system should be considered for point-to-point transmission. For underwater transmission through cables, HVDC is almost always the preferred choice. Lately, for stability reasons, back-to-back HVDC systems (without any transmission lines) are built for transferring of power between two AC systems where an AC-link will not be stable. More such back-to-back systems are expected to be used in the future.

7.2 POWER SEMICONDUCTOR DEVICES AND THEIR CAPABILITIES

These applications are made feasible by power semiconductor devices [1], whose symbols are shown in Figure 7.1a. The power handling capabilities and switching speeds of these devices are indicated in Figure 7.1b.

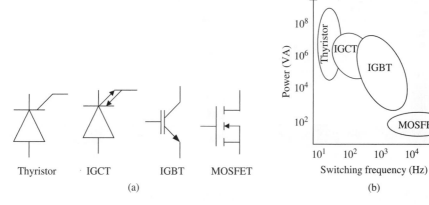

FIGURE 7.1 Power semiconductor devices.

All these devices allow current flow only in their forward direction (the intrinsic anti-parallel diode of MOSFETs can be explained separately). Transistors (intrinsically or by design) can block only the forward polarity voltage, whereas thyristors block both forward and reverse polarity voltages. Diodes are uncontrolled devices, which conduct current in the forward direction and block a reverse voltage. At very high power levels, integrated-gate controlled thyristors (IGCTs), which have evolved from the gate-turn-off thyristors, are sometimes used. Thyristors are semi-controlled devices that can switch-on at the desired instant in their forward-blocking state, but cannot be switched off from their gate and hence rely on the circuit in which they are connected to switch them off. However, thyristors are available in very large voltage and current ratings. Figure 7.2a shows the voltage and current ratings of various high power semiconductor devices in use, and Figure 7.2b shows the voltage and current ratings needed in various applications [2]. There is a great deal of research being carried out in SiC-based devices which are highly suited for applications at high power in power systems.

7.3   HVDC TRANSMISSION SYSTEMS

HVDC systems are represented by a one-line diagram of Figure 7.3 where the AC voltage at the sending-end is stepped up and supplied to a converter acting as a *rectifier* that converts AC into DC, and the power is transmitted over an HVDC transmission line. At the receiving end, there is another converter acting as an *inverter* that converters DC into AC that can be stepped down to the voltage level of the receiving-end system. The direction of the power flow can be reversible, thus reversing the role of the two converters. Figure 7.4 shows two types of HVDC systems: (a) a current-link system that uses thyristors and (b) a voltage-link system that uses switches such as IGBTs. Both of these systems are discussed in this chapter.

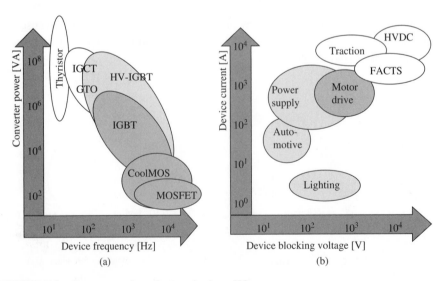

FIGURE 7.2   Power semiconductor devices [2].

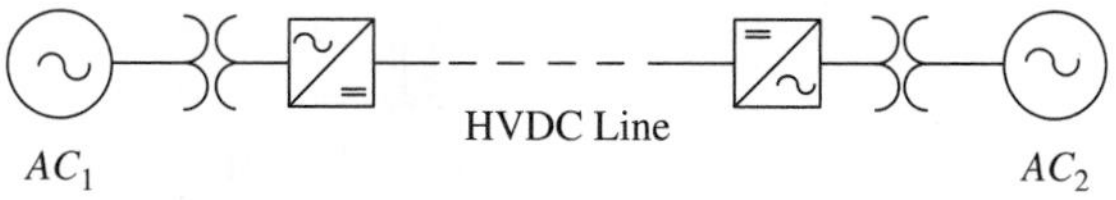

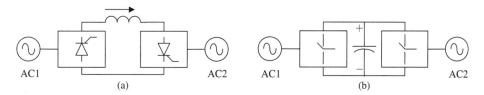

FIGURE 7.3   HVDC system—one-line diagram.

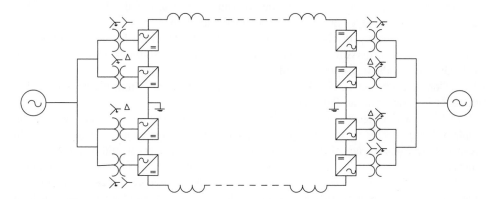

FIGURE 7.4   HVDC systems: (a) current-link and (b) voltage-link.

7.4   CURRENT-LINK HVDC SYSTEMS

The block diagram of a current-link HVDC system is shown in Figure 7.5. It consists of two poles: positive and negative with respect to ground. Each pole consists of two thyristors converters which are supplied through a $Y - Y$ and a $Y - \Delta$ arrangement of transformers to introduce a 30° phase shift between the two voltage sets, as discussed in the transformer chapter. In this current-link system, the transmission line inductance on the DC-side is usually supplemented by some extra inductance in series as shown in Figure 7.5 by the smoothing reactor. Since the current in the DC-link cannot change instantaneously because of these inductances, hence the name current-link. Each pole at the sending-end and the receiving-end consists of converters consisting of thyristors which are sometimes referred to by their trade name of silicone-controlled rectifiers (SCRs). The characteristic of these converters is explored further in the following subsection.

7.4.1   Thyristor Converters

We are familiar with diodes which block a negative polarity voltage and no current can flow in the reverse direction. Diodes begin to conduct current in the forward direction when a forward polarity voltage appears across them, with only a small voltage drop of

FIGURE 7.5   Block diagram of a current-link HVDC system.

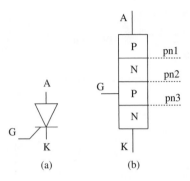

FIGURE 7.6 Thyristors.

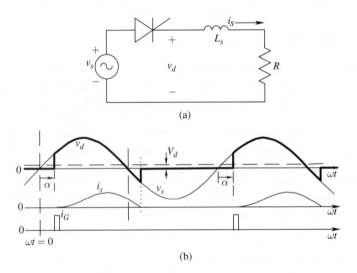

FIGURE 7.7 Thyristor circuit with a resistive load and a series inductance

the order a volt or two across them. These are represented by a symbol as shown in Figure 7.6a by an anode (A), a cathode (K), and a gate terminal (G). Unlike diodes, thyristors are a four-layer device, as shown in Figure 7.6b.

Similar to diodes, thyristors also conduct current only in the forward direction and block negative polarity voltage but unlike diodes, they can also block a forward-polarity voltage from conducting current as illustrated below.

Consider a primitive circuit to convert AC into DC as shown in Figure 7.7a with a thyristor in series with an R-L load.

As shown in Figure 7.7b, beginning at $\omega t = 0$ during the positive-half-cycle of the input voltage, a forward voltage appears across the thyristor (anode A is positive with respect to cathode K), and if the thyristor were a diode, a current would begin to flow in this circuit at $\omega t = 0$. This instant, at which current would begin to flow if it were a diode, is referred to as the instant of natural conduction. With the thyristor blocking the forward voltage, the start of conduction can be controlled (delayed) with respect to $\omega t = 0$ by a delay angle α at which instant a current pulse to the gate of the thyristor is applied. Once in the conducting state, the thyristor latches on and behaves like a diode with a very small

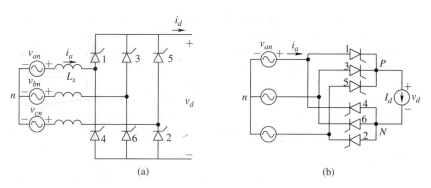

FIGURE 7.8 Three-phase full-bridge thyristor converter.

voltage drop of the order of 1 to 2 volts across it (we will idealize it as zero), and the *R-L* load voltage v_d equals v_s, as shown in Figure 7.7b.

The current waveform in Figure 7.7b shows that due to the inductor in series, the current through the thyristor keeps flowing for an interval into the negative half-cycle of the input voltage, coming to zero and staying at zero, not being able to reverse in direction due to the thyristor property, during the remainder of this half-cycle. In the next voltage cycle, the current conduction again depends on the instant during the positive half-cycle at which the gate pulse is applied. By controlling the delay angle (or the phase control as it is often referred), we can control the average voltage v_d across the *R-L* load. This principle can be extended to the practical circuits discussed below.

In HVDC systems, each converter uses six thyristors, as shown in Figure 7.8a. To explain the principle of converter operation, it is initially assumed that the converter is supplied by ideal three-phase voltage sources as shown in Figure 7.8b, thyristors are drawn by a top group and a bottom group, and the DC-side is represented by a DC current-source I_d.

Initially we will assume that the thyristors in Figure 7.8b are replaced by diodes. Diodes represent thyristor operation with the delay angle $\alpha = 0°$, and therefore replacing thyristors by diodes, we will be able to calculate the DC-side voltage for $\alpha = 0°$. Assuming diodes in Figure 7.8b, in the top group, all the diodes have their cathodes connected together and therefore only the diode with its anode connected to the highest phase voltage conducts and the other two become reverse-biased. In the bottom group, all the diodes have their anodes connected together, and therefore only the diode with its cathode connected to the lowest voltages conducts and the other two get reverse-biased. The waveforms at Points P and N, with respect to the source-neutral are shown in Figure 7.9a, where the DC-side voltage

$$v_d = v_{Pn} - v_{Nn} \tag{7.1}$$

is a line-line voltage within each 60° interval and is plotted in Figure 7.9b. The average DC-side voltage with $\alpha = 0$ can be calculated from the waveforms in Figure 7.9b by considering a 60° ($\pi/3$ rad) interval with which the waveforms repeat, where $\sqrt{2}\,V_{LL}$ is the peak value of the line-line input voltage as shown in Figure 7.9b:

$$V_{do} = \frac{1}{\pi/3} \int_{-\pi/6}^{\pi/6} \sqrt{2}V_{LL}\cos \omega t \cdot d(\omega t) = \frac{3\sqrt{2}}{\pi} V_{LL} \tag{7.2}$$

This average voltage is indicated in Figure 7.9b by a straight line as V_{do}, where the subscript "o" refers to $\alpha = 0$. The line current waveforms are shown in Figure 7.9c, where for example $i_a = I_d$ during the interval when phase-a has the highest voltage, and diode 1 connected to it is conducting. Similarly, $i_a = -I_d$ during the interval when phase-a has the lowest voltage, and diode 4 connected to it is conducting.

Delaying the gate pulses to the thyristors by an angle α measured with respect to their instants of natural conductions (the instant of natural conduction for a thyristor is the instant at which the current through it would begin to flow if α were to be zero), the waveforms are shown in Figure 7.10.

In the DC-side output voltage waveforms, the area A_α corresponds to "loss" in units of volt-radians due to delaying the gate pulses by α every $\pi/3$ radian. Assuming the time-origin in Figure 7.10 as the instant at which the phase voltage waveforms for phases a and

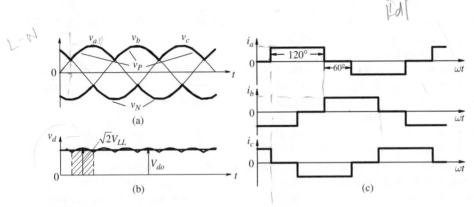

FIGURE 7.9   Waveforms in a three-phase rectifier with $L_s = 0$ and $\alpha = 0$.

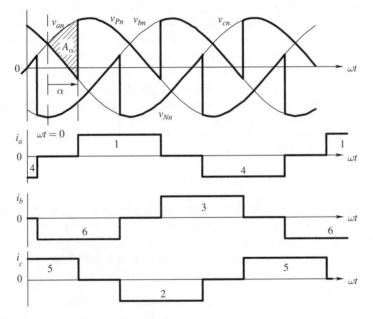

FIGURE 7.10   Waveforms with $L_s = 0$.

c cross, the line-line voltage v_{ac} waveform can be expressed as $\sqrt{2}V_{LL}\sin \omega t$. Therefore from Figure 7.10, the drop ΔV_α in the average DC-side voltage can be calculated as

$$\Delta V_\alpha = \frac{1}{\pi/3}\underbrace{\int_0^\alpha \sqrt{2}V_{LL}\sin \omega t \cdot d(\omega t)}_{A_\alpha} = \frac{3\sqrt{2}}{\pi}V_{LL}(1 - \cos \alpha) \qquad (7.3)$$

Therefore, using Equations 7.2 and 7.3, the average value of the DC-side voltage can be controlled by the delay angle as

$$V_{d\alpha} = V_{do} - \Delta V_\alpha = \frac{3\sqrt{2}}{\pi}V_{LL}\cos \alpha \qquad (7.4)$$

which, as shown by Equation 7.4, is positive for α between 0 and 90°—hence the converter is said to operate as a *rectifier*, whereas for α beyond 90°, $V_{d\alpha}$ becomes negative and the converter operates as an *inverter*.

Example 7.1
Three-phase thyristor converter of Figure 7.8b is operating in its inverter mode with $\alpha = 150°$. Draw waveforms similar to Figure 7.10 for this operating condition.

Solution   These waveforms for $\alpha = 150°$ in the inverter mode are shown in Figure 7.11. On the DC-side, the waveform of the voltage v_{Pn} is negative and that of v_{Nn} is positive.

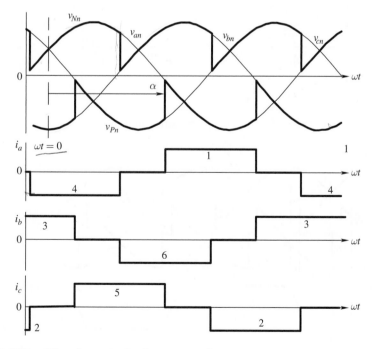

FIGURE 7.11   Waveforms in the inverter mode.

Therefore, the DC-side voltage $v_{PN}(= v_{Pn} - v_{Nn})$ and its average values are negative in the inverter-mode of operation. Since the DC-side current is in the same direction, the flow of power is from the DC-side to the AC-side. On the AC-side, phase-current waveforms are shifted (lagging) by $\alpha = 150°$, compared to the waveforms corresponding to $\alpha = 0$.

As shown in Figure 7.12a, for $\alpha < 90°$, the converter operates as a rectifier and the power flows, as shown in Figure 7.12b, from the AC-side to the DC-side. The opposite is true in the inverter mode with $\alpha > 90°$. In the inverter mode, the delay angle is limited to approximately 160° as shown in Figure 7.12a, due to the commutation angle required to safely commutate current from one thyristor to the next, as discussed below.

In the idealized case with $L_s = 0$, the AC-side currents commutate instantly from one thyristor to another, as shown by the current waveforms in Figure 7.10. However, in the presence of the AC-side inductance L_s, shown in Figure 7.8a, it takes a finite interval u during which the current commutates from one thyristor to another, as shown in Figure 7.13. During this commutation interval u, from α to $\alpha + u$, the instantaneous DC voltage is reduced due to the voltage drop v_L across the inductance in series with the thyristor to which the current is commutating from 0 to $(+I_d)$. Therefore, the average DC output voltage is reduced by an additional area A_u every $\pi/3$ radian as shown in Figure 7.13, where

$$A_u = \int_{\alpha}^{\alpha+u} v_L \, d(\omega t) = \omega L_s \int_0^{I_d} di_s = \omega L_s I_d \qquad (7.5)$$

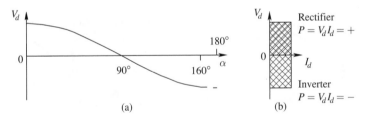

FIGURE 7.12   Average DC-side voltage as a function of α.

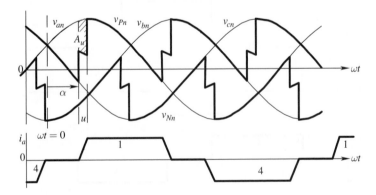

FIGURE 7.13   Waveforms with L_s.

and therefore an additional voltage drop due to the presence of L_s is

$$\Delta V_u = \frac{A_u}{\pi/3} = \frac{3}{\pi}\omega L_s I_d \tag{7.6}$$

Therefore, the DC-side output voltage can be written as

$$V_d = V_{d\alpha} - \Delta V_u \tag{7.7}$$

Substituting results from Equations 7.4 and 7.6 into Equation 7.7

$$V_d = \frac{3\sqrt{2}}{\pi} V_{LL}\cos\alpha - \frac{3}{\pi}\omega L_s I_d \tag{7.8}$$

In Figure 7.13, v_{Pn} during the commutation interval u is the average of v_{an} and v_{cn}. Therefore, Equation 7.8 can also be written as

$$V_d = \frac{3\sqrt{2}}{\pi} V_{LL}\cos(\alpha + u) + \frac{3}{\pi}\omega L_s I_d \tag{7.9}$$

Considering the AC-side of the converter, in Figure 7.13, the current for example in phase-a can be approximated as a trapezoid staring at an angle $(\pi/6 + \alpha)$ rads and rising linearly from $-I_d$ to $+I_d$ during the commutation angle u. With this approximation of a trapezoidal waveform, the fundamental component i_{a1} of the phase current lags behind the phase voltage by an angle $\phi_1(\simeq \alpha + u/2)$, as shown in Figure 7.14a in the rectifier mode, and in Figure 7.14b in the inverter mode. The power factor, always lagging, is as follows:

$$\text{Power Factor (PF)} \simeq \cos(\alpha + u/2) \tag{7.10}$$

The three-phase reactive power consumed by the converter is

$$Q_{3\phi} \simeq 3V_a I_{a1}\sin(\alpha + u/2) \tag{7.11}$$

In Equation 7.11, for approximate calculations, the AC-side current waveforms can be assumed to be rectangular (that is, $u = 0$), in which case $\hat{I}_{a1} = \frac{\sqrt{12}}{\pi}I_d$.

Avoiding Commutation Failure: The commutation of current from one thyristor to the new incoming thyristor is facilitated by the line-line AC voltage corresponding

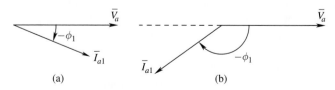

FIGURE 7.14   Power-factor angle.

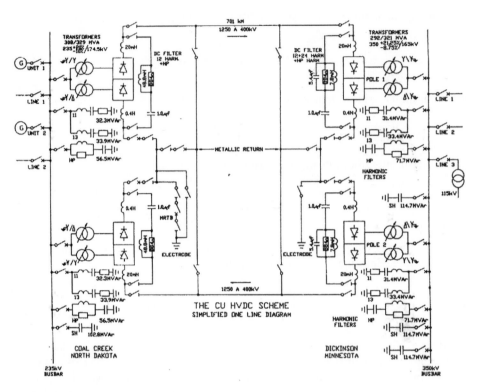

FIGURE 7.15 HVDC CU project [7].

to them, called the commutating voltage. In the inverter mode, if α is too large, then $(\alpha + u)$ may exceed 180 degrees. However, beyond 180 degrees, the polarity of the commutating voltage reverses and the current commutation is not successful. Therefore, to avoid this commutation failure, conservatively, α is limited to 160 degrees or so.

In the waveforms of Figure 7.10, there are six pulses, each with an interval of $60°(\pi/3 \text{ rad})$, every line-frequency cycle, and hence the converters for which these waveforms are drawn are called six-pulse converters. In most HVDC converter, in each pole, a 30° phase-shift is introduced by the $Y - Y$ and the $Y - \Delta$ transformer arrangements, as shown in Figure 7.5 and by the one-line diagram in Figure 7.15 of an actual HVDC system.

Each converter draws currents with a six-pulse waveform as shown in Figure 7.16a by $i_a(Y - Y)$ and $i_a(Y - \Delta)$. But the sum of these two currents, i_a, drawn from the utility, has a 12-pulse waveform that has much lower ripple content than the six-pulse waveform. Similarly, as shown in Figure 7.16b, the DC-side voltage waveforms of the two converters, v_{d1} and v_{d2}, in a pole sum up to a 12-pulse waveform v_d that has a much less ripple riding on the DC voltage, as compared to the six-pulse waveform. In spite of this reduction in ripple (or the harmonics, as they are called, of the fundamental line-frequency) filters are placed at the AC-side for the harmonics in currents and on the DC-side for the harmonics in the DC-side voltage, as shown in the one-line diagram of Figure 7.15 of an actual project, so that these harmonic currents and voltages do not interfere with the power system operation.

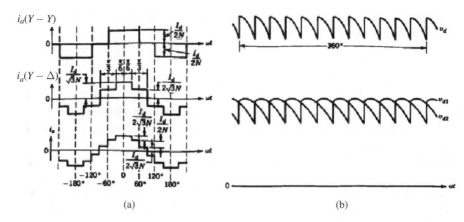

FIGURE 7.16   Six-pulse and 12-pulse current and voltage waveforms [3].

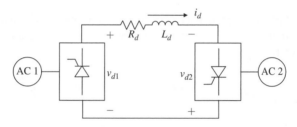

FIGURE 7.17   A pole of an HVDC system.

7.4.2   Power Flow in Current-Link Systems

Considering one of the poles, for example the positive pole in Figure 7.17, the two terminals of an HVDC system are shown in Figure 7.17a where R_d is the resistance of the DC-link inductance.

Since each terminal consists of two six-pulse thyristor converters as discussed previously,

$$V_{d1} = 2 \times \left[\frac{3\sqrt{2}}{\pi} V_{LL1}\cos\alpha_1 - \frac{3}{\pi}\omega L_{s1}I_d \right] \tag{7.12}$$

and

$$V_{d2} = 2 \times \left[\frac{3\sqrt{2}}{\pi} V_{LL2}\cos\alpha_2 - \frac{3}{\pi}\omega L_{s2}I_d \right] \tag{7.13}$$

where the AC-side voltages and the inductances may not be the same for the two terminals. Note that in Figure 7.17, the DC-side voltage of each terminal is defined as having a positive polarity where the current leaves. By controlling the delay angles α_1 and α_2 in a range from $0°$ to $160°$ as discussed previously, the average voltage, the average current, and the average power in the system of Figure 7.17 can be controlled, and the current through the DC-link is

$$I_d = \frac{V_{d1} + V_{d2}}{R_d} \tag{7.14}$$

where the DC-link resistance R_d is generally very small. In such a system, for the power flow from system 1 to system 2, V_{d2} is made negative by controlling α_2 such that it operates as an inverter and establishes the voltage of the DC-link. Converter 1 is operated as a rectifier at a delay angle α_1 such that it controls the current in the DC-link. The converse is true for these two converters if the power is to flow from system 2 to system 1. Generally in an inverter, the delay angle α is controlled so that the sum of the delay angle α and the commutation interval u remains constant and results in a constant Extinction Angle γ which is defined as follows:

$$\gamma = 180° - (\alpha + u) \tag{7.15}$$

This extinction angle is kept constant at a minimum value of usually 15° to 20°. Substituting Equation 7.15 into Equation 7.13 for converter-2 operating as an inverter, and making use of Equation 7.9

$$V_{d2} = 2 \times \left[-\frac{3\sqrt{2}}{\pi} V_{LL2} \cos \gamma_{\min} + \frac{3}{\pi} \omega L_{s2} I_d \right] \tag{7.16}$$

Using Equations 7.16 and 7.14

$$V_{d1} = 2 \times \frac{3\sqrt{2}}{\pi} V_{LL2} \cos \gamma_{\min} - \underbrace{\left(\frac{6}{\pi} \omega L_{s2} - R_d \right)}_{\text{positive}} I_d \tag{7.17}$$

The quantity within the "( )" in Equation 7.18 is generally positive. Therefore with converter 2 operating as an inverter at a constant minimum extinction angle $\gamma_{\min}$, the voltage V_{d1} given by Equation 7.17 is plotted in Figure 7.18 as a function of the DC-link current I_d. Converter 1 is operating as a rectifier with its delay angle controlled to maintain the DC-link current at its reference value $I_{d,ref}$. Therefore, its characteristic

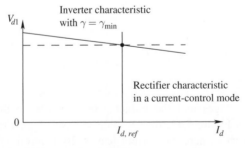

FIGURE 7.18 Control of an HVDC system [4].

appears as a vertical line in Figure 7.18. The intersection of the inverter and the rectifier characteristics establishes the operating point in terms of the voltage and the current in a HVDC system, as shown in Figure 7.18.

7.4.3   Improvements in Current-Link Systems

Recently, there are systems introduced which use active filters, compared to passive filters, to improve the system performance while reducing the space needed to install them. Also, there are systems introduced which consist of capacitors in series between the converter transformers and the converter bridges. These offer several advantages as described in [5-6].

7.5   VOLTAGE-LINK HVDC SYSTEMS

One of the limitations of the current-link system is that both the converters, regardless of their mode of operation, rectifier or inverter, require reactive power from the AC system. This limitation can be overcome in a voltage-link system. A voltage-link system shown earlier in Figure 7.4b is repeated in Figure 7.19a. Figure 7.19b shows the one-line diagram of one such system in operation [2].

In the block diagram of such a system shown in Figure 7.19a, each converter can independently either absorb or supply reactive power in a controllable manner. Unlike in current-link systems, in the voltage-link system, there is a capacitor on the DC-side of the converter in parallel that appears as a voltage port and hence the converters in such a system are also called voltage-source converters. One such converter between DC and three-phase AC is shown by its block diagram in Figure 7.20a. On a per-phase basis, this converter on the AC-side appears as a voltage source as shown in Figure 7.20b, interfacing the utility voltage source through a small inductance that may be the internal inductance of the utility system.

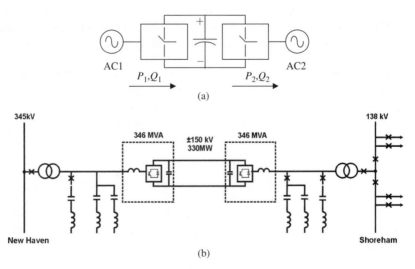

FIGURE 7.19   Block diagram of a voltage-link HVDC system [2].

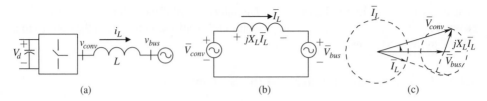

(a)   (b)   (c)

FIGURE 7.20 Block diagram of a voltage-link converter and the phasor diagram.

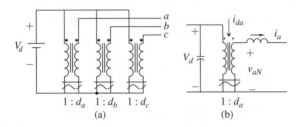

(a)   (b)

FIGURE 7.21 Synthesis of sinusoidal voltages.

The utility voltage can be assumed to be of constant amplitude with the reference phase angle of zero. Therefore, in the circuit of Figure 7.20b,

$$\overline{V}_{conv} = \overline{V}_{bus} + jX_L\overline{I}_L \tag{7.18}$$

The per-phase voltage synthesized by the converter can be controlled to be at the line-frequency, of the desired magnitude and phase where, as illustrated in Figure 7.20c, the tip of the $\overline{V}_{conv}$ phasor can be made to be at any point on the dotted circle. The resulting current $\overline{I}_L$ can thus be of the desired amplitude and phase, with the tip of $\overline{I}_L$ phasor on the dotted circle shown in Figure 7.20c. Hence, the power can be controlled in direction and magnitude, and the reactive power can be controlled in magnitude, absorbed or supplied, by the converter.

To synthesize these three-phase voltages, let us assume that we hypothetically have three ideal transformers available with continuously variable turns-ratios, as shown in Figure 7.21a. We will soon see that these ideal transformers are functional representations of the switch-mode converter that is required. We will focusing on only one of the three phases as shown in Figure 7.21b, others being identical in functionality, where from practical considerations, d_a is restricted to a range $0 \le d_a \le 1$. With this restriction, v_{aN} cannot become negative and therefore a DC offset of half the DC-bus voltage, $0.5\,V_d$, is introduced so that around this offset, the desired output voltage v_a, with respect to the output neutral, can become both positive and negative in a sinusoidal manner.

Therefore,

$$v_{aN} = 0.5V_d + \underbrace{\hat{V}_a\sin\omega t}_{v_a} \tag{7.19}$$

as shown in Figure 7.21. In this ideal transformer of Figure 7.21b, $v_{aN} = d_aV_d$ and hence the voltage in Equation 7.19 can be obtained by varying the turns-ratio $1 : d_a$ with time as follows, as shown in Figure 7.22

$$d_a = 0.5 + \hat{d}_a\sin\omega t \tag{7.20}$$

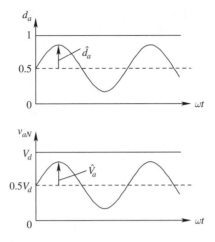

FIGURE 7.22 Sinusoidal variation of turns-ratio d_a.

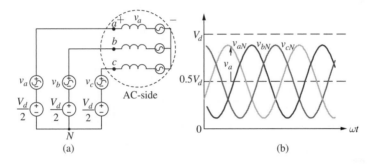

(a)   (b)

FIGURE 7.23 Three-phase synthesis.

where

$$\hat{V}_a = \hat{d}_a V_d \tag{7.21}$$

In Figure 7.23a, all three phase output voltages are shown and the waveforms for plotted in Figure 7.23b. In these outputs, the DC offsets (actually, these common-mode voltages need not be DC so long as they are the same in all three phases) are cancelled out from line-to-line voltages and hence can be ignored in consideration of the output voltages. It can be noted by Figure 7.23b that by introducing the common-mode voltages of $V_d/2$ in series with each phase, the maximum AC voltage magnitude is $\hat{V}_a = V_d/2$. By the so-called Space-Vector PWM (SV-PWM), as explained in [1], it is possible to modulate the common-mode voltage, rather than keeping it to $V_d/2$, and thus get the line-line voltage peak to equal V_d at the limit. This increases the output voltage capability of such a converter by approximately 15 percent.

Next we will study how this ideal transformer functionality is obtained. As shown in Figure 7.24a, a bi-positional switch within a two-port, a voltage-port on the DC-side and a current-port on the AC side, is used. This switch can be considered ideal, either up or down for a switching signal q_a of 1 or zero, respectively. In reality, such a switch can be constructed as shown in Figure 7.24b using two diodes and two IGBTs which are provided complementary gate signals, q_a and q_a^-.

In this bipositional switch, when $q_a = 1$ and the upper IGBT is on and the lower one off, and the current through the output inductor can flow in either direction, through the upper IGBT if positive or through the upper diode if negative. In either of these cases, the potential of point "a" is the same as that of the upper DC-bus and $v_{aN} = V_d$. Similarly, $q_a = 0$ results in $q_a^- = 1$ and the output current will flow either through the bottom diode or the bottom IGBT and hence $v_{aN} = 0$. Thus, the switching function q_a makes the switch operate as a bipositional switch, either up or down.

PWM. We will operate this switch at a high frequency, two or three orders of magnitude higher than the fundamental frequency to be synthesized. For example in synthesizing 60 Hz, the switching frequency f_s may be 6 kHz that is 100 times higher. This results in the output waveform shown in Figure 7.25a over one switching cycle $T_s(= 1/f_s)$, where the switching frequency f_s is kept constant. This output voltage has an

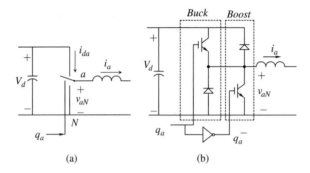

FIGURE 7.24   Realization of the ideal transformer functionality.

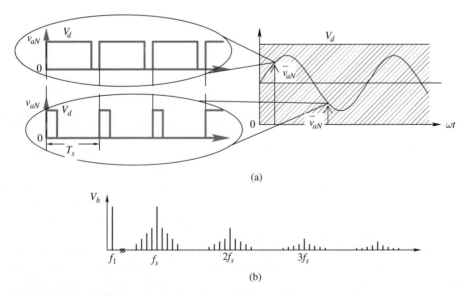

FIGURE 7.25   PWM to synthesize sinusoidal waveform.

average value, averaged over a switching-frequency time-period T_s, that can be written with a "-" on top, as

$$\bar{v}_{aN} = d_a V_d \tag{7.22}$$

where

$$d_a = \frac{T_{up}}{T_s} \tag{7.23}$$

In Equations 7.22 and 7.23, T_{up} and d_a can be continuously varied with time, sinusoidally with a DC offset, as discussed earlier, so that the average voltage given by Equation 7.22 is the same as that obtained by the ideal transformers in Figure 7.23, where v_{aN} and so on are ideal, without any ripple.

It should be recognized that in the output voltage, in addition to the desired average value Equation 7.22, there are unwanted switching harmonics as shown in Figure 7.25b at and around the multiples of the switching frequency, appearing as side bands at and around harmonic h, where

$$h = k_1 f_s \pm k_2 f_0 \tag{7.24}$$

The design of converters must take these harmonics into account and provide for adequate filtering so that they do not impact the AC system to which this converter is connected.

As explained earlier on a per-phase basis, as shown in Figure 7.20c, the current phasor can be controlled with respect to AC-system voltage, and thus the power flow can be in either direction, and the reactive power can be absorbed or supplied as needed. Thus in the block diagram of a voltage-link HVDC system in Figure 7.19a, ignoring losses, $P_1 = P_2$, but Q_1 and Q_2 are totally independent of each other in magnitude and direction, and each can be lagging or leading, and can help with the voltage stability as discussed in the stability chapter.

Use of the voltage-link HVDC converter concept is discussed for several other applications: in Chapter 3 in wind-turbine generators, in Chapter 8 for efficient control of motor speeds, and in Chapter 10 for reactive power control by FACTS devices such as STATCOM and UPFC for the voltage stability in power systems.

REFERENCES

1. N. Mohan, *Power Electronics—A First Course*, Wiley & Sons, 2011.

2. ABB Corporation (www.abb.com).

3. N. Mohan, T. Undeland, and W.P. Robbins, *Power Electronics: Converters, Applications, and Design*, 3rd edition, John Wiley & Sons, 2003.

4. E. W. Kimbark, Direct Current Transmission, vol. 1, Wiley–Interscience, New York, 1971.

5. G. Balzer, H. Müller, *Capacitor Commutated Converters for High Power HVDC Transmission*, Seventh International Conference on AC-DC Power Transmission, London, UK, 28–30 November 2001.

6. M. Meisingset, A. Golé, *A Comparison of Conventional and Capacitor Commutated Converters Based on Steady-State and Dynamic Considerations*, Seventh International Conference on AC-DC Power Transmission, London, UK, 28–30 November 2001.

7. Great River Energy (www.greatriverenergy.com).

PROBLEMS

7.1   In a three-phase thyristor converter, $V_{LL} = 460\,\text{V(rms)}$ at 60 Hz, and $L_s = 5\,\text{mH}$. The delay angle $\alpha = 30°$. This converter is supplying 5 kW of power. The DC-side current i_d can be assumed purely DC. (a) Calculate the commutation angle u, (b) draw the waveforms for the converter variables: phase voltages, phase currents, v_{Pn}, v_{Nn}, and v_d, and (c) assuming that the currents through the thyristors increase/decrease linearly during commutations, calculate the reactive power drawn by the converter.

7.2   In Figure 7.8a, assume $L_s = 0$ and $V_{LL} = 480\,\text{V}$ (RMS) at a frequency of 60 Hz. The delay angle $\alpha = 0°$. It is supplying a power of 10 kW. Calculate and plot the waveforms similar to those in Figure 7.10.

7.3   Repeat Problem 7.2 if the delay angle $\alpha = 45°$.

7.4   Repeat Problem 7.2 if the delay angle $\alpha = 145°$.

7.5   Repeat Problem 7.2 if in Problem 7.2 the AC-side inductance L_s is such that the commutation angle $u = 10°$.

7.6   Repeat Problem 7.2 if the delay angle $\alpha = 45°$ and the AC-side inductance L_s is such that the commutation angle $u = 10°$.

7.7   Repeat Problem 7.2 if the delay angle $\alpha = 145°$ and the AC-side inductance L_s is such that the commutation angle $u = 10°$. Note that the power flow is from the DC-side to the AC-side.

7.8   Calculate the reactive power consumed by the converter and the power factor angle in Problem 7.5.

7.9   Calculate the reactive power consumed by the converter and the power factor angle in Problem 7.6.

7.10   Calculate the reactive power consumed by the converter and the power factor angle in Problem 7.7. Note that the power flow is from the DC-side to the AC-side.

7.11   In the block diagram of Figure 7.17, for both three-phase converters, $V_{d0} = 480\,\text{kV}$. The DC-side current is $I_d = 1\,\text{kA}$. Converter 2 operating as an inverter establishes the DC-link voltage such that $V_{d2} = -425\,\text{kV}$. $R_d = 10.0\,\Omega$. The drop in DC voltage due to commutation overlap in each converter is 10 kV. In DC steady state, calculate the following angles: α_1, α_2, u_1 and u_2.

7.12   In Problem 7.11, calculate the reactive power drawn by each converter. Assume that the currents through the thyristors increase/decrease linearly during commutations.

7.13   In a voltage-link converter, the DC-bus voltage V_d and the three-phase voltages to be synthesized are given. (a) Write the expressions for v_{aN}, v_{bN}, and v_{cN}, and (b) write the expressions for d_a, d_b, and d_c

7.14   In Figure 7.20a, $\overline{V}_{bus} = 1\angle 0\,\text{pu}$ and $X_L = 0.1\,\text{pu}$. Calculate $\overline{V}_{conv}$ in order to supply 1 pu power at V_{bus} with (a) $Q = 0.5\,\text{pu}$, and (b) $Q = -0.5\,\text{pu}$.

PROBLEMS USING PSCAD/EMTDC

7.15   Analyze a primitive thyristor circuit, as described on the accompanying WEBSITE.

7.16   Obtain the waveforms in a six-pulse diode rectifier described on the WEBSITE accompanying this book.

7.17 Obtain the waveforms in a twelve-pulse diode rectifier described on the WEB-SITE accompanying this book.

7.18 Obtain the waveforms in a six-pulse thyristor converter operating in a rectifier-mode, as described on the accompanying WEBSITE.

7.19 Obtain the waveforms in a twelve-pulse thyristor converter operating in a recti-fier-mode, as described on the accompanying WEBSITE.

7.20 Obtain the waveforms in a six-pulse inverter, as described on the accompanying WEBSITE.

7.21 Obtain the waveforms in a twelve-pulse inverter, as described on the accompa-nying WEBSITE.

INCLUDING AN HVDC LINE IN POWER FLOW STUDIES USING *POWERWORLD*

7.22 Include an HVDC transmission system, as described on the accompanying WEBSITE, in the three-bus example power system.

8
DISTRIBUTION SYSTEM, LOADS, AND POWER QUALITY

8.1 INTRODUCTION

In this chapter, we will briefly examine the distribution system, the nature of prominent types of loads, and the power quality considerations to keep the power supply voltage sinusoidal at the rated voltage and line frequency.

8.2 DISTRIBUTION SYSTEMS

With the exception of distributed generation (DG) in the future, electricity today is normally generated remotely to the load centers and is transported through transmission lines at high, extra-high, and even ultra-high voltages. Electricity from this transmission network is passed on to the subtransmission network at voltages from 230 kV down to 35 kV. Some utilities consider this subtransmission network to be a part of the distribution system, while others consider only the network below 35 kV to be the distribution system. As stated in the chapter on transmission lines, approximately 9 percent of the electricity in the United States is dissipated in transmission and distribution systems, most of it in the later. Most industrial, commercial, and residential loads are supplied at much lower voltages than 34.5 kV, with the primary voltages in a range of 34.5 kV (19.92 kV per-phase) down to 12 kV (6.93 kV per-phase), and the secondary voltages at 480/277 V, three-phase, four-wire, and 120/240-V single-phase. Other secondary voltages are at 208/120 V, 480 V, and 600 V [1].

For residential loads, distribution substations supply single-phase distribution lines for various communities at voltages such as 13.8 kV. These voltages are stepped down locally to supply a set of houses at ±120 V as shown in Figure 8.1. At the entrance of a house, the neutral is grounded, and from the central breaker, a set of circuits, each with its own circuit breaker, are supplied by the +120 V line conductor and the neutral, whereas the other set is supplied by the −120 V line conductor and the neutral.

The ground conductor is carried along with each circuit to three-prong outlets, one for the line conductor, second for the neutral and the third for the ground conductor that is connected to the chassis of the load, for example a toaster. The reason for carrying along

132

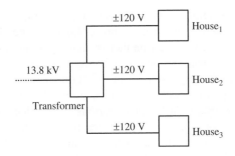

FIGURE 8.1 Residential distribution system.

the ground conductor is elimination of shock hazard. Normally there is no current flow through the ground conductor and it remains at ground potential, whereas the neutral conductor can rise above the ground potential due to current flow through its impedance. It should be noted that a current of only 5 mA is needed through the human heart to make it go into ventricular defibrillation. Outlets in more shock-susceptible areas with wet surfaces are equipped with ground fault interrupters (GFIs) that measure the difference between the current on the line conductor to that returning on the neutral conductor. The difference between the two currents indicates a fault to ground, and the GFI triggers the associated circuit breaker to be tripped.

8.3 POWER SYSTEM LOADS

Power systems are designed to serve industrial, commercial, and residential loads. A plot of the power demand on a utility, as a function of the time of day is plotted in Figure 8.2a as an example. The waveform of this load curve may be different during weekdays as compared to weekends that reflects shutdown of factories and commercial shops. The area underneath the plot in Figure 8.2a represents energy that the utility must supply in a 24-hour period, where as the peak of this waveform is the peak load that the utility must supply by its own generation or by purchasing power from other utilities. The ratio between the kilowatt-hours represented underneath the load curve in Figure 8.2a and the kilowatt-hours that would be necessary to generate if the load were constant at its peak value over the entire 24-hour period is called the *load factor*. The annual load-duration curve of Figure 8.2b [7] shows that the load is at or 90 percent above its peak value for only a small percentage of the time in a year.

Ideally, utilities would desire load factor to be unity but in reality it is far less than that. The reason for desiring the load factor of unity has to do with the fact that the peak

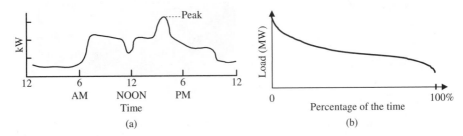

FIGURE 8.2 System load [7].

power is expensive to generate or to purchase from other utilities. It taxes the transmission system capacity and results in extra power losses. Some utilities have been very successful in increasing the load factor on their system by giving incentives to customers to shift their loads to off-peak periods by offering lower electricity rates at off-peak periods. Others have used energy storage in the form of pumped-hydro, for example, where, during off-peak periods, water is pumped from a lower reservoir up to a higher reservoir to generate electricity during peak periods. Somewhat along this line, utilities invest considerable sums of money in load forecasting over the next 24 hours to make purchasing agreements and to commit units to meet the expected load demand. Accurate forecasting of load results in considerable savings.

8.3.1   Nature of Power System Loads

Figure 8.3a shows the percentage of electricity consumed by the industrial, commercial and residential sectors, whereas Figure 8.3b shows electricity consumed by various types of loads in the United States.

Most utilities serve a variety of such loads. Industrial loads mainly depend on the type of industry, whereas the commercial and residential loads generally consist of the following in a variety of mixes:

- Electrical heating
- Lighting (incandescent and fluorescent)
- Motor loads to drive compressors for heating, ventilating, and air conditioning (HVAC)
- Power electronics based compressor loads and compact fluorescent lighting (CFL)

Each of these loads behaves differently for changes in the voltage magnitude and frequency. Generally, frequency does not vary appreciably in a large interconnected system such as that in North America (unless a phenomena called "islanding" occurs) to be concerned. However, the voltage sensitivity of loads must be included in calculating power flow and stability. Utilities estimate the mix of loads on their system, and knowing how each load behaves, the load aggregate is modeled in various studies by a combination of constant impedance, constant power and constant current representations.

Voltage sensitivities $a\,(=\partial P/\partial V)$ to real power and $b\,(=\partial Q/\partial V)$ to reactive power of various loads can be approximated as follows, and are summarized below in a tabular form:

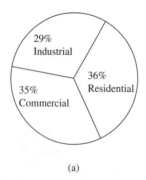

(a)

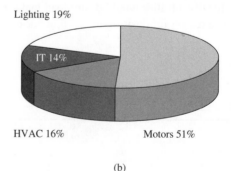

(b)

FIGURE 8.3   Utility loads.

- *Electric Heating*: These loads are resistive. Therefore their power factor equals unity; $a = 2$ and $b = 0$.
- *Incandescent Lighting*: Being resistive, their power factor is unity. Because the filament resistance is nonlinear, $a = 1.5$ and $b = 0$.
- *Fluorescent Lighting*: These use magnetic ballasts and their power factor is approximated at 0.9. It is reported that for such loads, $a = 1$ and $b = 1$.
- *Motor Loads*: Single-phase motors are used in smaller power ratings and three-phase motors in larger power ratings. Their power factor can be approximated in a range of 0.8 to 0.9. Their sensitivities to voltages depend of the type of load being driven, the fan-type or the compressor-type, because of the variation of the torque required by the load as a function of speed. Motor sensitivities are reported to be in a range of $a = 0.05 - 0.5$ and $b = 1 - 3$.
- *Modern Power Electronics−Based Loads*: Modern and future power electronics−based loads, that are increasingly being used, are nonlinear; they continue to draw the same power even if the input voltage were to change slightly in its magnitude and frequency, as discussed in this section. They can be designed with power factor−corrected (PFC) interface that results in essentially a unity power factor. In such loads, ideally, it is possibly to get $a = 0$ and $b = 0$. Structure of these loads is described in further detail below.

8.3.2 Power Electronics−Based Loads

The trend in power system loads is that they are increasingly being supplied through power electronics interface. Doing so increases the overall system efficiency, in some cases as much as 30 percent—for example in heat pump systems [2]—and hence has a large energy conservation potential. In most cases, these loads are supplied by a voltage-link system, discussed in connection with HVDC transmission systems, as shown in Figure 8.4.

TABLE 8.1 Approximate Power Factor and Voltage Sensitivity of Various Loads

Type of Load	Power Factor	$a = \partial P / \partial V$	$b = \partial Q / \partial V$
Electric Heating	1.0	2.0	0
Incandescent Lighting	1.0	1.5	0
Fluorescent Lighting	0.9	1.0	1.0
Motor Loads	0.8−0.9	0.05−0.5	1.0−3.0
Modern Power Electronics−Based Loads	1.0	0	0

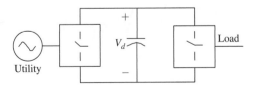

FIGURE 8.4 Voltage-link system for modern and future power electronics−based loads.

A great deal of lighting load is shifting to compact fluorescent lamps (CFLs) where a power-electronics interface shown in Figure 8.4 is needed to produce a high-frequency AC in the range of 30 kHz to 40 kHz at which these lamps operate most efficiently. CFLs, compared to incandescent lamps are approximately four times more efficient. That is, to provide the same illumination, they consume only one-fourth the electricity, hence amount to huge savings. In spite of their high initial cost, they are being used in increasing numbers, even in developing countries.

Figure 8.5 shows the per-phase steady state equivalent circuit of a three-phase induction motor. Conventionally, the speed ω_m of such motors has been controlled by reducing the magnitude V_a of the applied voltage, without changing its frequency, that is, by keeping the synchronous speed ω_{syn} unchanged. (Note that ω_{syn} in rad/s equals $(2/p)2\pi f$, where f is the frequency of the voltages in Hz and p is the number of poles. Therefore, ω_{syn} remains constant if f is not changed.) Operating at large values of slip speed ω_{slip}, which equals $(\omega_{syn} - \omega_m)$, causes large power losses in the rotor circuit, resulting in very low energy efficiency of operation.

Speed of induction-motor loads can be adjusted efficiently by a power electronics interface of Figure 8.4 that produces three-phase output voltages whose amplitude and frequency can be controlled, independent of each other. By controlling the frequency of the voltages being applied to an induction motor, provided the voltage magnitude is also controlled to result in the rated air gap flux within the motor, the motor torque-speed characteristics for various frequencies can be plotted as shown in Figure 8.6.

Each frequency of applied voltage results in the corresponding synchronous speed ω_{syn} from which the slip speed ω_{slip} is measured. Assuming a constant-torque load characteristic as shown dotted in Figure 8.6, the operating speed can be varied

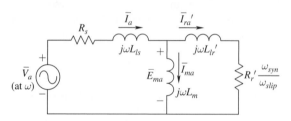

FIGURE 8.5 Per-phase, steady state equivalent circuit of a three-phase induction motor.

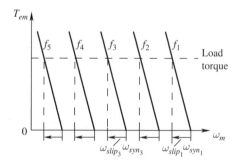

FIGURE 8.6 Torque-speed characteristic of induction motor at various applied frequencies.

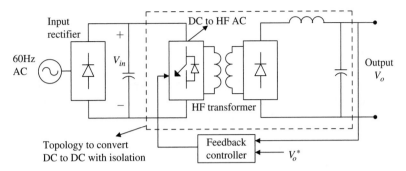

FIGURE 8.7   Switch-mode DC power supply.

continuously by continuously varying the applied frequency. At each operating point, the slip speed, measured with respect to its corresponding synchronous speed, remains small, and hence the energy efficiency remains high.

In applications mentioned earlier where a regulated DC power supply is needed, the voltage link system of Figure 8.4 is used, as shown by the block diagram of Figure 8.7, to produce high-frequency AC, which is stepped down in voltage by a high-frequency transformer and then rectified to produce a regulated DC output. The transformer operates at a high frequency in a range of 200 kHz to 300 kHz and hence can be made much smaller than a line-frequency transformer of the similar VA rating. Efficiencies of such switch-mode DC power supplies approach 90 percent, which is almost double of those associated with linear power supplies.

In developed countries with high cost of electricity, most of the compressor-loads such as air-conditioners and heat pumps are supplied through a power electronics interface of the type shown in Figure 8.4. Studies have shown that by adjusting the compressor speed to match the thermal load, as much as 30 percent savings in electricity can be achieved, as compared to the conventional on/off cycling approach. Similarly, in pump-driven systems to adjust flow-rate, controlling the pump speed by using the power-electronics interface of Figure 8.4, considerable improvement in the overall system efficiency can be obtained as compared to the use of a throttling valve. A report by DOE [3] estimates that if all the pump-driven systems in the United States were controlled using power electronics, a mature technology now, enough energy can be saved annually to equal that used by the state of New York!

Power electronics based loads are often nonlinear in the sense they often continue to draw the same power irrespective of the change in the input voltage in a small range. If these power electronics based loads are not designed appropriately, they draw distorted (non-sinusoidal) currents from the utility source and hence can degrade the quality of power as discussed in the next section. However, it is possible to design power electronics interface of Figure 8.4 with current shaping circuits, often called power-factor-correction (PFC) circuits [4], so that the current drawn from the utility is sinusoidal and at a unity power factor.

8.4   POWER QUALITY CONSIDERATIONS

It is important to customers that the power they receive from their utility is of acceptable quality. These power quality considerations can be classified into the following categories:

- Continuity of Service
- Magnitude of Voltage
- Voltage Waveform

In an interconnected system such as that in North America, the frequency of voltage supply is seldom of concern and therefore is not mentioned above. We will discuss each of the above considerations in the following subsections.

8.4.1   Continuity of Service

Most serious power quality issue is the lack of the continuity of service. Utilities do their best to ensure continuity because a service disruption also means a loss of revenue for them. Interconnected systems, as one of their benefits, improve the continuity of service. If a part of the power system goes down for whatever reason, an interconnected system has a better chance of supplying power through an alternative route.

8.4.1.1   Uninterruptible Power Supplies (UPS)
To further improve this continuity for critical loads such as some computers and medical equipment, uninterruptible power supplies (UPS) are used that store energy in chemical batteries, and also in flywheels in the form of inertial energy, within the voltage-link system shown in Figure 8.8.

8.4.1.2   Solid State Transfer Switches
On a larger scale for an entire industrial plant itself, it is possible to make use of two feeders if they are derived from different systems, as shown in Figure 8.9. One of the feeders is used as the primary feeder that normally supplies power to the load. In case of a fault on this feeder, solid state transfer switches, made up of thyristors or IGBTs, switch the load to the alternate feeder within a few milliseconds and hence maintain the continuity of service. This action, of course, assumes that the same fault, in all probability far away, does not affect both feeders.

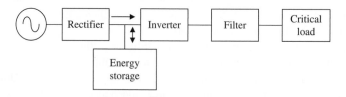

FIGURE 8.8   Uninterruptible power supply.

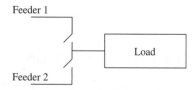

FIGURE 8.9   Alternate feeder.

8.4.2 Voltage Magnitude

Power system loads prefer that the voltage magnitude be at its specified nominal value.

Utilities try to maintain voltage supply within a range such as ±5 percent of the nominal voltage. However fault conditions, even far away, and heavy or very light loading of transmission lines can cause voltages to be outside the normal range in one or more phases. To maintain this voltage in magnitude, equipment such as dynamic voltage restorers (DVRs) is used. As shown in the block diagram form in Figure 8.10, utilizing a voltage link system, DVRs inject a voltage in series with the utility supply to bring the voltage at the consumer-end to within the acceptable range.

On a larger scale, in distribution substations, voltage regulators and transformers with load-tap changing (LTC) automatically try to regulate the supply voltage. According to [5], both three-phase and single-phase voltage regulators are used in distribution substations to regulate the load-side voltage. Substation regulators are one of the primary means, along with load-tap-changing power transformers, shunt capacitors, and distribution line regulators, for maintaining a proper level of voltage at a customer's service entrance. A very important function of substation voltage regulation is to correct for supply voltage variation. With the proper use of the control settings and line drop compensation, regulators can correct for load variations as well. A properly applied and controlled voltage regulator not only keeps the voltage at a customer's service entrance within approved limits but also minimizes the range of voltage swing between light and heavy load periods [5].

This problem of voltage outside the normal range can be mitigated by reactive power control using a power-electronics based device such as a Static Compensator, known as STATCOM, shown in Figure 8.11 as a block diagram. A STATCOM acts as a continually adjustable reactor, which can draw either inductive or capacitive volt-amperes (vars) within its designed rating, to regulate the bus voltage. STATCOMs are further explained in Chapter 10 dealing with the voltage stability.

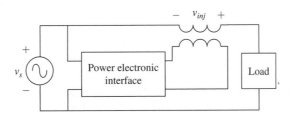

FIGURE 8.10 Dynamic voltage restorer (DVR).

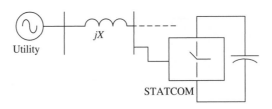

FIGURE 8.11 STATCOM [4].

8.4.3 Voltage Waveform

Most of the load equipment in use is designed assuming a sinusoidal voltage waveform. Any distortion in this voltage waveform can cause loads such as induction motors to draw distorted (non-sinusoidal) currents which can result in loss of efficiency and overheating and may cause some loads to fail. As mentioned earlier, power-electronics based loads, if not designed with this consideration, would draw distorted currents and would cause the supply voltage, due to its internal impedance, to become distorted and hence cause problems to other adjacent loads.

8.4.3.1 Distortion and Power Factor

To quantify distortion in the current drawn by power electronic systems, it is necessary to define certain indices.

As a base case, consider the linear $R - L$ load shown in Figure 8.12a which is supplied by a sinusoidal source in steady state. The voltage and current phasors are shown in Figure 8.12b, where ϕ is the angle by which the current lags the voltage.

The voltage and current phasors are shown in Figure 8.12b, where ϕ is the angle by which the current lags the voltage. Using RMS values for the voltage and current magnitudes, the average power supplied by the source is

$$P = V_s\, I_s \cos \phi \tag{8.1}$$

The power factor (PF) at which power is drawn is defined as the ratio of the real average power P to the product of the RMS voltage and the RMS current:

$$PF = \frac{P}{V_s I_s} = \cos \phi \quad \text{(using Equation 8.1)} \tag{8.2}$$

where $V_s I_s$ is the apparent power. For a given voltage, from Equation 8.2, the RMS current drawn is

$$I_s = \frac{P}{V_s \cdot PF} \tag{8.3}$$

This shows that the power factor PF and the current I_s are inversely proportional. The current flows through the utility distribution lines, transformers, and so on, causing losses in their resistances. This is the reason why utilities prefer unity power factor loads that draw power at the minimum value of the RMS current.

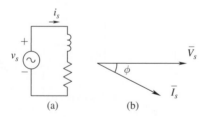

(a)   (b)

FIGURE 8.12   Voltage and current phasors in simple *R-L* circuit.

8.4.3.2 RMS Value of Distorted Current and the Total Harmonic Distortion (*THD*)

The sinusoidal current drawn by the linear load in Figure 8.12 has zero distortion. However, power electronic systems without a power-factor-corrected (PFC) front-end draw currents with a distorted waveform such as that shown by $i_s(t)$ in Figure 8.13a. The utility voltage $v_s(t)$ is assumed sinusoidal. The following analysis is general, applying to the utility supply that is either single-phase or three-phase, in which case the analysis is on a per-phase basis.

The current waveform $i_s(t)$ in Figure 8.13a repeats with a time-period T_1. By Fourier analysis of this repetitive waveform, we can compute its fundamental frequency $(= 1/T_1)$ component $i_{s1}(t)$, shown dotted in Figure 8.13a. The distortion component $i_{distortion}(t)$ in the input current is the difference between $i_s(t)$ and the fundamental-frequency component $i_{s1}(t)$:

$$i_{distortion}(t) = i_s(t) - i_{s1}(t) \tag{8.4}$$

where $i_{distortion}(t)$ using Equation 8.4 is plotted in Figure 8.13b. This distortion component consists of components at frequencies that are the multiples of the fundamental frequency.

To obtain the RMS value of $i_s(t)$ in Figure 8.13a, we will apply the basic definition of RMS:

$$I_s = \sqrt{\frac{1}{T_1} \int_{T_1} i_s^2(t) \cdot dt} \tag{8.5}$$

Using Equation 8.4,

$$i_s^2(t) = i_{s1}^2(t) + i_{distortion}^2(t) + 2i_{s1}(t) \times i_{distortion}(t) \tag{8.6}$$

In a repetitive waveform, the integral of the products of the two harmonic components (including the fundamental) at unequal frequencies, over the repetition time-period, equals zero:

$$\int_{T_1} g_{h_1}(t) \cdot g_{h_2}(t) \cdot dt = 0 \qquad h_1 \neq h_2 \tag{8.7}$$

Therefore, substituting Equation 8.6 into Equation 8.5, and making use of Equation 8.7 that implies that the integral of the third term on the right side of Equation 8.6 equals zero,

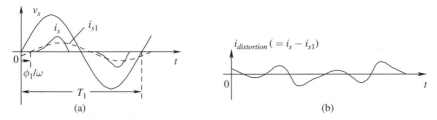

(a)   (b)

FIGURE 8.13 Current drawn by power electronics equipment without PFC.

$$I_s = \sqrt{\underbrace{\frac{1}{T_1}\int_{T_1} i_{s1}^2(t)\cdot dt}_{I_{s1}^2} + \underbrace{\frac{1}{T_1}\int_{T_1} i_{distortion}^2(t)\cdot dt}_{I_{s1}^2} + 0} \qquad (8.8)$$

or

$$I_s = \sqrt{I_{s1}^2 + I_{distortion}^2} \qquad (8.9)$$

where the RMS values of the fundamental-frequency component and the distortion component are as follows:

$$I_{s1} = \sqrt{\frac{1}{T_1}\int_{T_1} i_{s1}^2(t)\cdot dt} \qquad (8.10)$$

and

$$I_{distortion} = \sqrt{\frac{1}{T_1}\int_{T_1} i_{distortion}^2(t)\cdot dt} \qquad (8.11)$$

Based on the RMS values of the fundamental and the distortion components in the input current $i_s(t)$, a distortion index called the total harmonic distortion (*THD*) is defined in percentage as follows:

$$\% \, THD = 100 \times \frac{I_{distortion}}{I_{s1}} \qquad (8.12)$$

Using Equation 8.9 into Equation 8.12,

$$\% \, THD = 100 \times \frac{\sqrt{I_s^2 - I_{s1}^2}}{I_{s1}} \qquad (8.13)$$

The RMS value of the distortion component can be obtained based on the harmonic components (except the fundamental) as follows using Equation 8.7:

$$I_{distortion} = \sqrt{\sum_{h=2}^{\infty} I_{sh}^2} \qquad (8.14)$$

where I_{sh} is the rms value of the harmonic component h.

8.4.3.3 Obtaining Harmonic Components by Fourier Analysis

By Fourier analysis, any distorted (non-sinusoidal) waveform $g(t)$ that is repetitive with a fundamental frequency f_1, for example i_s in Figure 8.13a, can be expressed as a sum of sinusoidal components at the fundament and its multiples (harmonic) frequencies:

$$g(t) = G_0 + \sum_{h=1}^{\infty} g_h(t) = G_0 + \sum_{h=1}^{\infty} \{a_h \cos(h\omega t) + b_h \sin(h\omega t)\} \tag{8.15}$$

where the average value G_0 is DC

$$G_0 = \frac{1}{2\pi} \int_0^{2\pi} g(t) \cdot d(\omega t) \tag{8.16}$$

The sinusoidal waveforms in Equation 8.15 at the fundamental frequency f_1 ($h = 1$) and the harmonic components at frequencies h times f_1 can be expressed as the sum of their cosine and sine components

$$a_h = \frac{1}{\pi} \int_0^{2\pi} g(t) \cos(h\omega t) d(\omega t) \qquad h = 1, 2, \ldots, \infty \tag{8.17}$$

$$b_h = \frac{1}{\pi} \int_0^{2\pi} g(t) \sin(h\omega t) d(\omega t) \qquad h = 1, 2, \ldots, \infty \tag{8.18}$$

The cosine and the sine components above, given by Equations 8.17 and 8.18, can be combined and written as a phasor in terms of its RMS value

$$\overline{G}_h = G_h \angle \phi_h \tag{8.19}$$

where the RMS magnitude in terms of the peak values a_h and b_h equals

$$G_h = \frac{\sqrt{a_h^2 + b_h^2}}{\sqrt{2}} \tag{8.20}$$

and the phase ϕ_h can be expressed as

$$\tan \phi_h = \frac{-b_h}{a_h} \tag{8.21}$$

It can be shown that the rms value of the distorted function $g(t)$ can be expressed in terms of its average and the sinusoidal components as

$$G = \sqrt{G_0^2 + \sum_{h=1}^{\infty} G_h^2} \tag{8.22}$$

In Fourier analysis, by appropriate selection of the time origin, it is often possible to make the sine or the cosine components in Equation 8.15, thus considerably simplifying the analysis, as illustrated by a simple example.

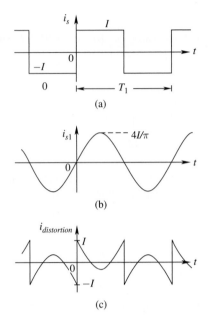

FIGURE 8.14   Example 8.1.

Example 8.1
A current i_s of square waveform is shown in Figure 8.14a. Calculate and plot its fundamental frequency component and its distortion component. What is the *%THD* associated with this waveform?

Solution   From Fourier analysis, by choosing the time origin as shown in Figure 8.14a, $i_s(t)$ in Figure 8.14a can be expressed as

$$i_s = \frac{4}{\pi}I\left(\sin \omega_1 t + \frac{1}{3}\sin 3\omega_1 t + \frac{1}{5}\sin 5\omega_1 t + \frac{1}{7}\sin 7\omega_1 t + \ldots \right) \qquad (8.23)$$

The fundamental frequency component and the distortion component are plotted in Figures 8.14b and 8.14c.

From Figure 8.14a, it is obvious that the RMS value I_s of the square waveform is equal to I. In the Fourier expression of Equation 8.23, and the RMS value of the fundamental-frequency component is

$$I_{s1} = \frac{(4/\pi)}{\sqrt{2}}I = 0.9I$$

Therefore, the distortion component can be calculated from Equation 8.9 as

$$I_{distortion} = \sqrt{I_s^2 - I_{s1}^2} = \sqrt{I^2 - (0.9I)^2} = 0.436I$$

Therefore, using the definition of *THD*,

$$\%THD = 100 \times \frac{I_{distortion}}{I_{s1}} = 100 \times \frac{0.436I}{0.9I} = 48.4\%$$

8.4.3.4 The Displacement Power Factor (*DPF*) and Power Factor (*PF*)

Next, we will consider the power factor at which power is drawn by a load with a distorted current waveform such as that shown in Figure 8.13a. As before, it is reasonable to assume that the utility-supplied line-frequency voltage $v_s(t)$ is sinusoidal, with an RMS value of V_s and a frequency $f_1(= \frac{\omega_1}{2\pi})$. Based on Equation 8.7, which states that the product of the cross-frequency terms has a zero average, the average power P drawn by the load in Figure 8.13a is due only to the fundamental-frequency component of the current:

$$P = \frac{1}{T_1} \int_{T_1} v_s(t) \cdot i_s(t) \cdot dt = \frac{1}{T_1} \int_{T_1} v_s(t) \cdot i_{s1}(t) \cdot dt \tag{8.24}$$

Therefore, in a load that draws distorted current

$$P = V_s I_{s1} \cos \phi_1 \tag{8.25}$$

where ϕ_1 is the angle by which the fundamental-frequency current component $i_{s1}(t)$ lags behind the voltage, as shown in Figure 8.13a.

At this point, another term called the displacement power factor (*DPF*) needs to be introduced, where

$$DPF = \cos \phi_1 \tag{8.26}$$

Therefore, using the *DPF* in Equation 8.25,

$$P = V_s I_{s1}(DPF) \tag{8.27}$$

In the presence of distortion in the current, the meaning and therefore the definition of the power factor, at which the real average power P is drawn, remains the same as in Equation 8.2—that is, the ratio of the real power to the product of the RMS voltage and the RMS current:

$$PF = \frac{P}{V_s I_s} \tag{8.28}$$

Substituting Equation 8.27 for P into Equation 8.28,

$$PF = \left(\frac{I_{s1}}{I_s}\right)(DPF) \tag{8.29}$$

In linear loads that draw sinusoidal currents, the current-ratio (I_{s1}/I_s) in Equation 8.29 is unity, hence $PF = DPF$. Equation 8.29 shows the following: a high distortion in the current waveform leads to a low power factor, even if the *DPF* is high. Using

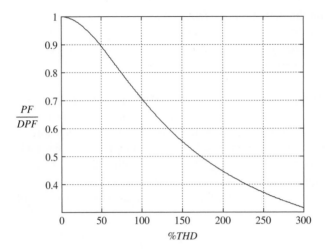

FIGURE 8.15   Relation between PF/DPF and THD.

Equation 8.13, the ratio (I_{s1}/I_s) in Equation 8.29 can be expressed in terms of the total harmonic distortion as

$$\frac{I_{s1}}{I_s} = \frac{1}{\sqrt{1 + \left(\frac{\%THD}{100}\right)^2}} \tag{8.30}$$

Therefore, in Equation 8.29,

$$PF = \frac{1}{\sqrt{1 + \left(\frac{\%THD}{100}\right)^2}} \cdot DPF \tag{8.31}$$

The effect of *THD* on the power factor is shown in Figure 8.15 by plotting (PF/DPF) versus *THD*. It shows that even if the displacement power factor is unity, a total harmonic distortion of 100 percent (which is possible in power electronic systems unless corrective measures are taken) can reduce the power factor to approximately 0.7 ($\frac{1}{\sqrt{2}} = 0.707$ to be exact), which is unacceptably low.

8.4.3.5   Deleterious Effects of Harmonic Distortion and a Poor Power Factor
There are several deleterious effects of high distortion in the current waveform and the poor power factor that results due to it.

- Power loss in utility equipment such as distribution and transmission lines, transformers, and generators increases, possibly to the point of overloading them.
- Harmonic currents can overload the shunt capacitors used by utilities for voltage support and may cause resonance conditions between the capacitive reactance of these capacitors and the inductive reactance of the distribution and transmission lines.
- The utility voltage waveform will also become distorted, adversely affecting other linear loads, if a significant portion of the load supplied by the utility draws power by means of distorted currents.

In order to prevent degradation in power quality, recommended guidelines (in the form of the IEEE-519) have been suggested by the IEEE (Institute of Electrical and Electronics Engineers). These guidelines place the responsibilities of maintaining power quality on the consumers and the utilities as follows: (1) on the power consumers, such as the users of power electronic systems, to limit the distortion in the current drawn, and (2) on the utilities to ensure that the voltage supply is sinusoidal with less than a specified amount of distortion.

The limits on current distortion placed by the IEEE-519 are shown in Table 8.2, where the limits on harmonic currents, as a ratio of the fundamental component, are specified for various harmonic frequencies. Also, the limits on the *THD* are specified. These limits are selected to prevent distortion in the voltage waveform of the utility supply.

Therefore, the limits on distortion in Table 8.2 depend on the "stiffness" of the utility supply, which is shown in Figure 8.16a by a voltage source $\overline{V}_s$ in series with internal impedance Z_s. An ideal voltage supply has zero internal impedance. In contrast, the voltage supply at the end of a long distribution line, for example, will have large internal impedance. To define the "stiffness" of the supply, the short-circuit current I_{sc} is calculated by hypothetically placing short-circuit at the supply terminals, as shown in Figure 8.16b. The stiffness of the supply must be calculated in relation to the load current. Therefore, the stiffness is defined by a ratio called the short-circuit ratio *(SCR)*:

$$\text{Short-Circuit Ratio } SCR = \frac{I_{sc}}{I_{s1}} \qquad (8.32)$$

where I_{s1} is the fundamental-frequency component of the load current. Table 8.2 shows that a smaller short-circuit ratio corresponds to lower limits on the allowed distortion in the current drawn. For the short-circuit ratio of less than 20, the total harmonic distortion in the current must be less than 5 percent. Power electronic systems that meet this limit would also meet the limits of more stiff supplies.

TABLE 8.2 Harmonic current distortion (I_h/I_1)

	Odd Harmonic Order h (in %)					Total Harmonic
I_{sc}/I_1	$h < 11$	$11 \le h < 17$	$17 \le h < 23$	$23 \le h < 25$	$35 \le h$	Distortion(%)
< 20	4.0	2.0	1.5	0.6	0.3	5.0
$20-50$	7.0	3.5	2.5	1.0	0.5	8.0
$50-100$	10.0	4.5	4.0	1.5	0.7	12.0
$100-1000$	12.0	5.5	5.0	2.0	1.0	15.0
> 1000	15.0	7.0	6.0	2.5	1.4	20.0

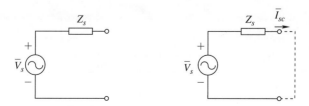

FIGURE 8.16 (a) Utility supply, (b) Short-circuit current.

It should be noted that the IEEE-519 does not propose harmonic guidelines for individual pieces of equipment but rather for the aggregate of loads (such as in an industrial plant) seen from the service entrance, which is also the point-of-common-coupling (PCC) with other customers. However, the IEEE-519 is frequently interpreted as the harmonic guidelines for specifying individual pieces of equipment such as motor drives. There are other harmonic standards, such as the IEC-1000, which apply to individual pieces of equipment.

8.4.3.6 Active Filters

Harmonic currents produced by power electronics loads can be prevented from entering the utility system by means of filters. These filters are often passive filters tuned to certain harmonic frequencies, for example, in HVDC terminals. Lately, active filters have also been employed where a current is produced by power electronics means and is injected into the utility system such as to nullify the harmonic currents produced by the nonlinear load. Therefore, only a sinusoidal current is drawn from the utility source by the combination of the nonlinear load and the active filter.

8.5   LOAD MANAGEMENT [6,7] AND SMART GRID

This is a topic that is likely to become extremely important in coming years as the utilities are stretched to meet the load demand. The load management can take many forms. Utilities can implement time-of-day rates, thus giving customers incentives to shift their loads to off-peak hours. They can implement demand-side management (DMS) where certain loads, such as air conditioners, can be interrupted remotely during peak-load hours and in return, customers who sign up for it, get rebates on their electricity bills. Large customers

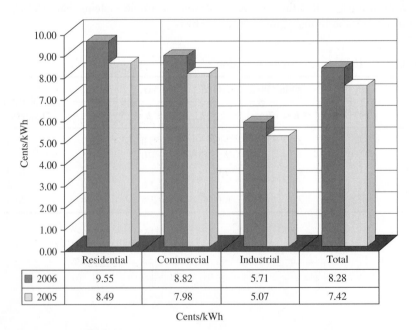

	Residential	Commercial	Industrial	Total
■ 2006	9.55	8.82	5.71	8.28
☐ 2005	8.49	7.98	5.07	7.42

Cents/kWh

FIGURE 8.17   Average retail price of electricity to ultimate customers [3].

can often negotiate to pay a reduced rate for the energy (kWh) used, in addition to a demand charge based on the peak power (kW) they draw in a given month. Load shedding based on voltage and frequency can be an important strategy for maintaining proper system operation and preventing voltage collapse and blackouts. Smart grids using smart meters can provide incentives to consumers to draw power based on the time-of-day prices.

8.6 PRICE OF ELECTRICITY [3]

It is informative to know the average retail price of electricity to ultimate customers by the end-use sector and it has changed recently, as shown in Figure 8.17. These prices keep changing from year to year.

REFERENCES

1. IEEE Std 141-1986.
2. N. Mohan and J. W. Ramsey, *Comparative Study of Adjustable-Speed Drives for Heat Pumps*, EPRI Report, 1986.
3. U.S. Department of Energy (www.eia.doe.gov).
4. N. Mohan, *Power Electronics: A First Course*, Wiley & Sons, 2011 (www.wiley.com).
5. United States Department of Agriculture, Rural Utilities Service, Design Guide for Rural Substations, RUS BULLETIN 1724E-300 (http://www.rurdev.usda.gov/RDU_Bulletins_Electric.html).
6. J. Casazza and F. Delea, *Understanding Electric Power Systems: An Overview of the Technology and the Marketplace*, IEEE Press and Wiley-Interscience, 2003.
7. Electric Utility Systems and Practices, 4th edition, Homer M. Rustebakke (editor), John Wiley & Sons, 1983.

PROBLEMS

8.1 Describe voltage distribution inside homes and residential buildings.

8.2 What is the role of ground fault interrupters (GFIs), and how do they work?

8.3 What is meant by the load factor in describing the daily load curve on utility systems?

8.4 What is a typical make up of utility loads and what are their real and reactive power sensitivities to voltages?

8.5 What is the characteristic of power electronics based loads; explain.

8.6 A load in steady state is characterized by $P = 1$ pu and $Q = 0.5$ pu at a voltage $V = 1$ pu. Represent it as a constant-impedance load.

8.7 What are the main power quality considerations?

8.8 Describe in words the nature of the CBEMA curve.

8.9 What are interruptible power supplies?

8.10 What is meant by the dual feeder arrangement?

8.11 What are dynamic voltage restorers (DVRs), and how do they work?

8.12 What means are used in substations to regulate voltages?

8.13 What are STATCOMs, and how do they work?

8.14 How is the total harmonic distortion defined in current waveforms?

8.15 What is meant by power-factor-correction (PFC) circuits?

8.16   How do active filters work?

8.17   In a single-phase power electronics load, $I_s = 10$ A (rms), $I_{s1} = 8$ A (rms), and $DPF = 0.9$. Calculate $I_{distortion}$, %THD, and PF.

8.18   In a single-phase power electronics load, the following operating conditions are given: $V_s = 120$ V(rms), $P = 1$ kW, $I_{s1} = 10$ A, and $THD = 80$ %. Calculate the following: DPF, $I_{distortion}$, I_s, and PF.

8.19   What are active filters, and how do they work?

8.20   Define the following terms and their significance: load management, demand-side management, load shedding, demand charges, time-of-day rates, load forecasting, and annual load-duration curve.

PROBLEMS USING PSCAD/EMTDC

8.21   Calculate the displacement power factor, power factor and the total harmonic distortion associated with the power-electronics interface, as described on the accompanying website.

9

SYNCHRONOUS GENERATORS

9.1 INTRODUCTION

In most power plants, hydro, steam, or gas turbines provide mechanical input to synchronous generators that convert mechanical input to three-phase electrical power output. Synchronous generators, by means of their voltage excitation system, called the voltage regulator, are also the primary means of supporting and regulating the system voltage to its nominal value. Thousands of such generators operate in synchronism in the interconnected system in North America. In this chapter, we will briefly look at the basic structure of synchronous generators and the fundamental principles of the electromagnetic interactions that govern their operation. These synchronous generators can be classified into two broad categories:

- Turbo alternators used with steam turbines, as shown in Figure 9.1a, or with gas turbines that rotate at high speeds such as at 1,800 rpm for producing the 60-Hz output (or at 1,500 rpm for the 50-Hz output)
- Hydro generators used in hydro power plants, as shown in Figure 9.1b, where turbines rotate at very slow speeds (few hundred rpm), hence these generators are very large with a large number of poles to produce the output at the line-frequency of 60 or 50 Hz

We will focus on the high-speed turbo-alternators of Figure 9.1a, although the basic principles involved also apply to the hydro generators. Such generators can be several hundred MVA in ratings at voltages around 20 kV.

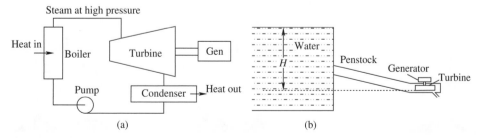

FIGURE 9.1 Synchronous generators driven by (a) steam turbines and (b) hydraulic turbines.

9.2   STRUCTURE

Electrical generators are designed to be long and cylindrical as shown in Figure 9.2a. To discuss their characteristics, we will describe them by their cross-section by hypothetically slicing them by a plane perpendicular to the shaft axis in Figure 9.2b by a plane. Looking from one side, Figure 9.2b shows the stationary part, called the stator, made up of magnetic material such as silicon steel and firmly affixed to the foundation. The rotor on a set of bearings is free to rotate, and the stator and the rotor are separated by a very small air gap.

Winding placed on the stator and the rotor produce magnetic flux; Figure 9.3 shows the representation of flux lines in various types of generators. Figures 9.3a and b show a round rotor (non-salient) generator, where the generator is a two-pole generator in Figure 9.3a and a four-pole generator in Figure 9.3b. The cross-section in Figure 9.3c illustrates a four-pole salient pole generator where there are distinct poles, called salient poles, on the rotor. Hydro generators may consist of dozens of such pole-pairs in a salient construction where the magnetic reluctance to flux lines is much smaller in the radial path through the rotor-pole axis as compared to a path that is in between the two adjacent poles.

In multi-pole generators with more than one pole-pair, as shown in Figures 9.3b and c, it is sufficient to consider only one pole-pair consisting of adjacent north and south poles due to complete symmetry around the periphery in the air gap. Other pole-pairs have identical conditions of magnetic fields and currents. Therefore, a multi-pole generator (with the number of poles $p > 2$) can be analyzed by considering that a pole-pair spans 2π electrical radians and expressing the rotor speed in the units of electrical radians per second to be $p/2$ times its mechanical speed. For the ease of explanation in this chapter, we will assume a two-pole generator with $p = 2$.

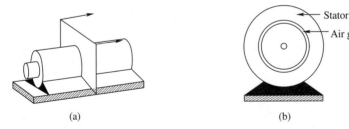

(a)   (b)

FIGURE 9.2   Machine cross-section.

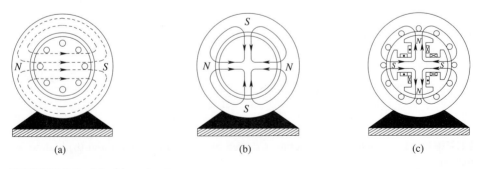

(a)   (b)   (c)

FIGURE 9.3   Machine structure.

In order to minimize the ampere-turns required to create flux lines shown crossing the air gap in Figure 9.3, both the rotor and the stator consist of high permeability ferromagnetic materials and the length of the air gap is kept as small as possible, which is shown highly exaggerated for the ease of drawing. As in transformers, to reduce eddy-current losses, the stator consists of silicon-steel laminations, which are insulated from each other by a layer of thin varnish. These laminations are stacked together, perpendicular to the shaft axis. Conductors, which run parallel to the shaft axis, are placed in the slots cut into these laminations. Hydrogen, oil, or water is often used as the coolant.

9.2.1 Stator with Three-Phase Windings

In synchronous generators, the stator has three-phase windings, with their respective magnetic axes, as shown in Figure 9.4a. The windings for each phase ideally should produce a sinusoidally distributed flux density in the air gap in the radial direction. Theoretically, this requires a sinusoidally distributed winding in each phase. In practice, this is approximated in a variety of ways. To visualize this sinusoidal distribution, consider the winding for phase *a*, shown in Figure 9.4b, where, in the slots, the number of turns-per-coil for phase-*a* progressively increases away from the magnetic axis, reaching

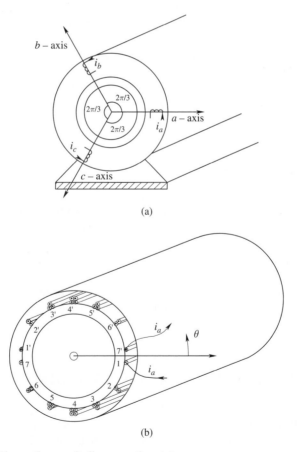

FIGURE 9.4 Three phase windings on the stator.

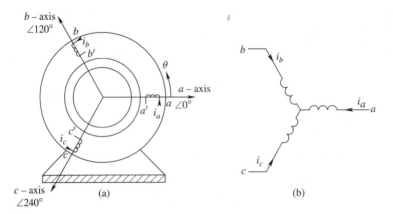

FIGURE 9.5   Connection of three phase windings.

a maximum at $\theta = 90°$. Each coil, such as the coil with sides 1 and 1′, spans 180 degrees where the current into coil-side 1 returns in 1′ through the end-turn at the back of the generator. This coil (1,1′) is connected in series to coil-side 2 of the next coil (2,2′), and so on. Graphically, these windings are simply drawn as shown in Figure 9.4a with the understanding that each is sinusoidally distributed, with its magnetic axis as shown.

In Figure 9.4b, we focused only on phase-a, which has its magnetic axis along $\theta = 0°$. There are two more identical sinusoidally-distributed windings for phases b and c, with magnetic axes along $\theta = 120°$ and $\theta = 240°$, respectively, as represented in Figure 9.5a. These three windings are generally connected in a wye-arrangement by connecting terminals $a′$, $b′$, and $c′$ together, as shown in Figure 9.5b. Flux-density distributions in the air gap due to currents i_b and i_c, identical in shape to that due to i_a, peak along their respective phase-b and phase-c magnetic axes.

9.2.2   Rotor with DC Field Winding

Rotor of a synchronous generator contains a field winding in its slots. This winding is supplied by a DC voltage, resulting in a DC current I_f. The field-current I_f in Figure 9.6 produces the rotor field in the air gap. It is desired that this field flux-density be sinusoidally distributed in the air gap in the radial direction, effectively producing north and south poles, as shown in Figure 9.6. By controlling I_f and hence the rotor-produced field, it is possible to control the induced emf of this generator, and control the reactive power delivered by it. This field-excitation system is discussed later in this chapter.

In steady state, the rotor rotates at a speed, called the synchronous-speed ω_{syn} in rad/s, which in a two-pole machine equals $\omega(= 2\pi f)$, where f is the frequency of the generated output voltages.

9.3   INDUCED EMF IN THE STATOR WINDINGS

To discuss induced emfs, we will concentrate on phase-a, realizing that similar emfs are induced in the other two phase windings. The phase-a winding is shown in Figure 9.7 by a single coil where it is understood that this winding is sinusoidally distributed as discussed earlier. Current i_a in the direction shown in Figure 9.7 is so chosen that it results in

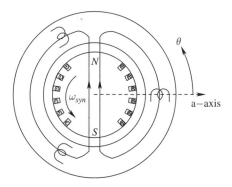

FIGURE 9.6   Field winding on the rotor that is supplied by a DC current I_f.

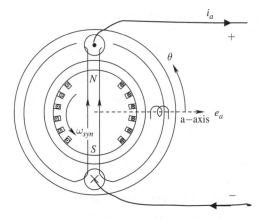

FIGURE 9.7   Current direction and voltage polarity; the rotor position shown induces maximum e_a.

flux-lines that peak along the phase-a magnetic axis. The induced emf polarity in Figure 9.7 is chosen to be positive where the current exits, since we are interested in the generator-mode of operation (as opposed to the motoring mode, where following the passive sign-convention would require that the current enter the positive terminal of the induced voltage).

In the stator windings, there are two causes of induced emfs, which will be discussed one at a time, and assuming that there is no magnetic saturation, these two induced emfs will be superimposed to yield the resultant induced emf.

9.3.1   Induced EMF due to Rotation of the Field-Flux with the Rotor

In Figure 9.8a, at time $t = 0$ is chosen such that the axis of the field winding is vertically up. Due to this field winding, the density of the flux-lines in ϕ_f that links the stator is distributed co-sinusoidally in the radial direction, peaking along the axis of the field-winding. This field-flux density distribution can be represented by a space vector $\vec{B}_f$ as shown in Figure 9.8b at time $t = 0$. The length of the space vector represents the value of the flux-density peak, and its angle measured with respect to the phase-a axis represents

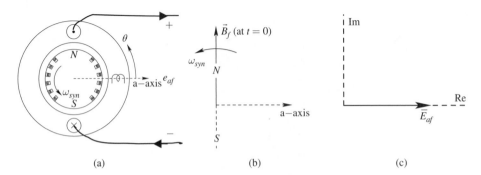

FIGURE 9.8 Induced emf e_{af} due to rotating rotor-field with the rotor.

its orientation. This space vector is distinguished from phasors by an "arrow" on top. As the rotor rotates, so does the $\vec{B}_f$ space vector.

Due to the rotor field-flux lines rotating with the rotor and cutting the stationary windings in the stator, voltages are induced in the stator phase-windings in accordance with Faraday's Law. At an instant when the rotor and the $\vec{B}_f$ orientation are as shown in Figures 9.8a and b respectively, and rotating at a speed ω_{syn}, the voltage induced at this instant will be maximum in phase-a since the highest density of field-flux lines are cutting the highest density of phase-a conductors. If the rotor is rotating counterclockwise, then by Lenz's Law, we can determine the induced voltage in phase-a at $t = 0$ to be positive with the voltage-polarity defined in Figure 9.8a. As the rotor rotates with time, the induced emf in phase-a will vary co-sinusoidally with time; this voltage is represented by a phasor $\overline{E}_{af}$, as shown in Figure 9.8c.

Both, the space vector $\vec{B}_f$ in the space vector diagram of Figure 9.8b and the phasor $\overline{E}_{af}$ in the phasor diagram of Figure 9.8c are complex variables—that is, they both have amplitudes and angles. Therefore, from Figures 9.8b and c, these two can be related as follows:

$$\overline{E}_{af} = (-j)k_f\vec{B}_f(0) \tag{9.1}$$

where $\vec{B}_f(0)$ is the space vector at time $t = 0$, k_f is a constant of proportionality depending on the machine construction details and the synchronous speed ω_{syn}. The reason for $(-j)$ in Equation 9.1 is that $\overline{E}_{af}$ lags behind $\vec{B}_f(0)$ by 90 degrees.

9.3.2 Induced EMF due to the Rotating Magnetic Field Called the Armature Reaction, Created by the Stator Currents

When a generator is connected to the electrical grid, the result is a flow of sinusoidal phase currents. These currents are necessary to produce power that is supplied to the grid. Flow of these phase currents produces a rotating magnetic field that, in addition to the rotating field-flux, also "cuts" the stator phase windings that are stationary.

In Figure 9.9a, each phase current results in a pulsating flux-density distribution that peaks along its phase axis, and is proportional to instantaneous value of the phase current; at any instant, the flux-density drops off co-sinusoidally away from its phase axis. Therefore, the flux-density distribution produced by each phase winding can be represented by a space vector, oriented along the respective phase axis. Each of these space vectors is stationary in position but its amplitude pulsates with time as the phase-current value changes with time, and can be expressed as

$$\vec{B}_{i_a} = (k_1 i_a)e^{j0} \qquad \vec{B}_{i_b} = (k_1 i_b)e^{j2\pi/3} \qquad \vec{B}_{i_c} = (k_1 i_c)e^{j4\pi/3} \tag{9.2}$$

where k_1 is a machine constant that relates the instantaneous phase-current and the peak flux-density values. The resultant flux-density distribution can be obtained by vectorially summing the three space vectors, using the principle of superposition by assuming a linear magnetic circuit:

$$\vec{B}_{AR} = k_1 \left(i_a e^{j0} + i_b e^{j2\pi/3} + i_c e^{j4\pi/3} \right) \tag{9.3}$$

where the subscript AR refers to the armature reaction, as it is commonly called.

Let the three phase currents be as below, where $t = 0$ corresponds to the rotor position as shown in Figure 9.8a, $\omega\, (= 2\pi f)$ is the frequency of the stator voltages and currents, and θ is an angle by which the phase-current lags the generated voltage $\overline{E}_{af}$ in Figure 9.8c:

$$i_a = I_a \cos(\omega t - \theta) \qquad i_b = I_a \cos(\omega t - \theta - 2\pi/3) \qquad i_c = I_a \cos(\omega t - \theta - 4\pi/3) \tag{9.4}$$

where I_a is the RMS value of each phase current. The sinusoidal current i_a is represented by a phasor $\overline{I}_a$ in Figure 9.9b; similarly other phase currents can be represented as phasors. Note that in steady state,

$$\omega = \omega_{syn} \tag{9.5}$$

By substituting the current expressions of Equation 9.4 into Equation 9.3, and making use of Equation 9.5,

$$\vec{B}_{AR} = (k_2 I_a)e^{j(\omega_{syn}t - \theta)} \tag{9.6}$$

where recalling that $\vec{B}_{AR}$ is the space vector that represents the flux-density distribution due to all three phase currents at any instant t. In steady state, the peak of this space vector remains constant (k_2 is another constant) for a given current rms magnitude I_a, and

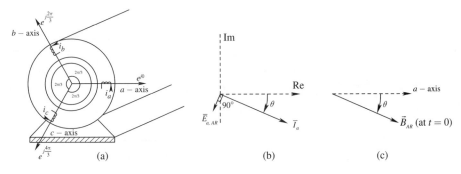

FIGURE 9.9 Armature reaction due to phase currents.

it rotates counterclockwise at the synchronous speed ω_{syn} with time. From Equations 9.4 and 9.6, we should note that the orientation of $\vec{B}_{AR}$ at time $t = 0$ in the space-vector diagram of Figure 9.9c is the same as the orientation of $\bar{I}_a$ in the phasor diagram of Figure 9.9b.

Just like the rotating field-flux density distribution $\vec{B}_f$ induces field voltages $\bar{E}_{af}$ in the stationary stator phase-a winding, rotating $\vec{B}_{AR}$ induces armature-reaction voltages $\vec{E}_{a,AR}$; similar voltages are induced in the other two phases b and c, time-delayed by 120 degrees and 240 degrees, respectively. Therefore, making the analogy with Equation 9.1, $\bar{E}_{a,AR}$ shown in Figure 9.9b lags behind $\vec{B}_{AR}(0)$ shown in Figure 9.9c by 90°, and it can be expressed as follows:

$$\bar{E}_{a,AR} = (-j)k_3\vec{B}_{AR}(0) \tag{9.7}$$

where k_3 is another constant of proportionality. Figures 9.9b and c show that the orientation of $\vec{B}_{AR}(0)$ at $t = 0$ in the space-vector diagram is the same as that of $\bar{I}_a$ in the phasor diagram, and the magnitude of both these variables are related through Equations 9.6 and 9.7. Therefore, the armature-reaction voltage in phase-a can be written as

$$\bar{E}_{a,AR} = -jX_m\bar{I}_a \tag{9.8}$$

Equation 9.8 shows that $\bar{E}_{a,AR}$ lags behind $\bar{I}_a$ by 90°, and the magnitude of these two are related to each other by what is called the magnetizing reactance X_m of the synchronous generator.

9.3.3 Combined Induced EMFs due to the Field-Flux and the Armature Reaction

We saw in the above two sub-sections that the induced emf in phase-a (and similarly in the other phases b and c) is due to two mechanisms. In a magnetic circuit, assuming no saturation, these two emfs can be combined to determine the resultant emf:

$$\bar{E}_a = \bar{E}_{af} + \bar{E}_{a,AR} = \bar{E}_{af} - jX_m\bar{I}_a \tag{9.9}$$

as shown in Figure 9.10a, where $\bar{I}_a$ lags $\bar{E}_{af}$ by the same angle θ as defined in Equation 9.4. The relationship of Equation 9.9 can be represented by a per-phase equivalent shown

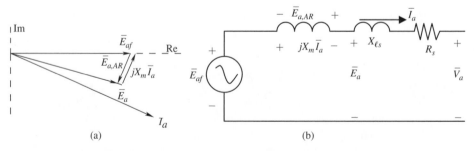

(a)   (b)

FIGURE 9.10 Phasor diagram and per-phase equivalent circuit.

in Figure 9.10b. Including the effect of the leakage flux by a voltage drop across the leakage reactance $X_{\ell s}$, and including the voltage drop across the phase-winding resistance R_s, we can write the terminal voltage expression as

$$\overline{V} = \overline{E}_{af} - jX_s\overline{I}_a - R_s\overline{I}_a \qquad (9.10)$$

where $X_s(= X_{\ell s} + X_m)$ is called the synchronous reactance, which is the sum of the leakage reactance $X_{\ell s}$ of each stator winding and the magnetizing reactance X_m.

9.4 POWER OUTPUT, STABILITY, AND THE LOSS OF SYNCHRONISM

Induced emfs in stator windings cause phase currents to flow, which produce an electromechanical torque that opposes the torque supplied by the turbine. In the circuit of Figure 9.11a, consider a generator connected to an infinite bus (an ideal voltage source) $\overline{V}_\infty$ through a radial line.

The reactance X_T is the sum of the generator synchronous reactance and the internal reactance of the utility grid (plus the leakage reactance of transformer(s) if any). Choosing $\overline{V}_\infty$ as the reference phasor (i.e., $\overline{V}_\infty = V_\infty \angle 0$) and neglecting the circuit resistance in comparison to the reactance, the power from the fundamental concepts in Chapter 3 can be written for all three phases as

$$P = 3\frac{E_{af}V_\infty}{X_T}\sin\delta \qquad (9.11)$$

where the rotor angle δ associated with $\overline{E}_{af}(= E_{af}\angle\delta)$ is positive in the generator mode; the rotor angle δ is a measure of the angular displacement or position of the rotor with respect to a synchronously rotating reference axis.

If the field current is kept constant, then the magnitude E_{af} is also constant in steady state, and thus the power output of the generator is proportional to the sine of the torque angle δ between $\overline{E}_{af}$ and $\overline{V}_\infty$. This power-angle relationship is plotted in Figure 9.11b for both positive and negative values of δ, where for negative values of δ, this machine goes into its motoring mode.

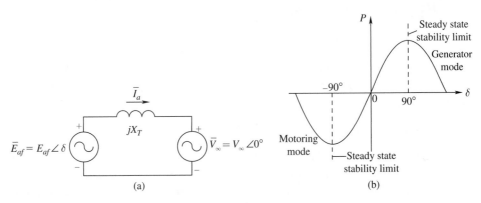

FIGURE 9.11 Power output and synchronism.

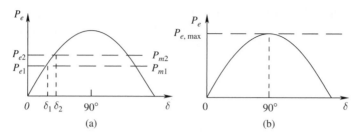

FIGURE 9.12   Steady state stability limit.

9.4.1   Steady State Stability Limit

Figure 9.11b shows that the power supplied by the synchronous generator reaches its peak at $\delta = 90°$. This is the steady-state limit, beyond which the synchronism is lost. This can be explained as follows: Initially, assuming no losses, at a value δ_1 below 90 degrees, the turbine is supplying power P_{m1} that equals the electrical output P_{e1}, as shown in Figure 9.12a. To supply more power, the power input from the turbine is increased (for example, by letting more steam into the turbine). This momentarily speeds up the rotor, causing the torque angle δ associated with the rotor-induced voltage $\overline{E}_{af}$ to increase. Finally, a new steady state is reached at $P_{e2}(= P_{m2})$ a higher value of the torque angle δ_2 as shown in Figure 9.12a.

However, at and beyond $\delta = 90°$, if the power input from the turbine is increased as shown in Figure 9.12b, then increasing δ causes the electrical output power to decline, which results in a further increase in δ (because more mechanical power is coming in, while less electrical power is going out). This increasing δ causes an intolerable increase in generator currents and the protection relays cause the circuit breakers to trip to isolate the generator from the grid, thus saving the generator from being damaged.

The above sequence of events is called the "loss of synchronism," and the stability is lost. In practice, the transient stability due to a sudden change in the electrical power output, forces the maximum value of the steady-state torque angle δ to be much less than 90 degrees, typically in a range of 40 to 45 degrees.

9.5   FIELD EXCITATION CONTROL TO ADJUST REACTIVE POWER

The reactive power associated with synchronous generators can be controlled in magnitude as well as in sign (leading or lagging). To discuss this, let us assume, as a base case, that a synchronous generator is supplying a constant power, and the field current I_f is adjusted such that this power is supplied at a unity power factor, as shown in the phasor diagram of Figure 9.13a.

9.5.1   Overexcitation

Now, an increase in the field current, called overexcitation, will result in a larger magnitude of $\overline{E}_{af}$, since assuming no magnetic saturation, E_f depends linearly on the field-current I_f. However, $E_{af}\sin\delta$ must remain constant from Equation 9.11, since the power output is constant. Similarly, the projection of the current phasor $\overline{I}_a$ on the voltage phasor

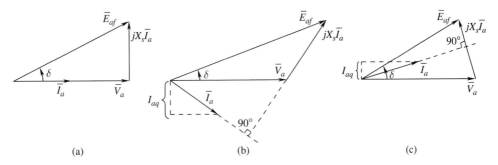

FIGURE 9.13 Excitation control to supply reactive power (neglecting R_s).

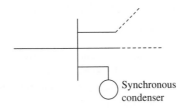

FIGURE 9.14 Synchronous Condenser.

$\overline{V}_a$ must remain the same as in Figure 9.13a. These result in the phasor diagram of Figure 9.13b, where the current $\overline{I}_a$ is lagging behind $\overline{V}_a$. Considering the utility grid to be a load, the grid absorbs the reactive power as an inductor does. Therefore, a synchronous generator operating in an overexcited mode supplies reactive power, like a capacitor does. The three-phase reactive power Q can be computed from the reactive component of the current I_{aq} as

$$Q = 3V_a\, I_{aq} \tag{9.12}$$

9.5.2 Underexcitation

In contrast to the overexcitation, decreasing I_f results in a smaller magnitude E_{af}, and the corresponding phasor diagram, assuming that the power output remains constant as before, can be represented as in Figure 9.13c. Now the current $\overline{I}_a$ leads the voltage $\overline{V}_a$, and the load (the utility grid) supplies reactive power, as a capacitor does. Thus, a synchronous generator in the underexcited mode absorbs reactive power like an inductor does. The three-phase reactive power Q can be computed from the reactive component of the current I_{aq}, similar to Equation 9.12.

9.5.3 Synchronous Condensers

Sometimes in power systems, grid-connected synchronous machines are operated in a motoring mode to supply reactive power, as shown in Figure 9.14. A similar control over the reactive power in these synchronous condensers, as they are often called, can be exercised by controlling the field excitation as explained above. There is no need for a turbine to operate these machines and the small amount of power loss accrued in operating the synchronous machine as a motor is supplied by the grid.

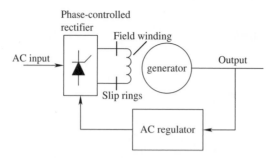

FIGURE 9.15   Field exciter for automatic voltage regulation (AVR).

9.6   FIELD EXCITERS FOR AUTOMATIC VOLTAGE REGULATION (AVR)

Field excitation of synchronous generators can be controlled to regulate the voltage at their terminals or at some other bus in the system for that matter, usually to its nominal value. This is possible since the voltage regulation and the supply of reactive power are related, and the objective of regulating the voltage at a designated bus dictates what reactive power the generator should supply. Most generators are equipped with an automatic voltage regulator that senses the bus voltage to be regulated and compares it with its desired value. The error between the two is calculated within the regulator shown in Figure 9.15, which by means of the phase-controlled rectifier, controls the DC voltage applied to the field-excitation winding to adjust the field current I_f appropriately.

These field-excitation systems can take several forms based on where the input power is derived from, and the desire to avoid slip-ring and brushes that are necessitated because the field winding to be supplied is rotating with the rotor.

9.7   SYNCHRONOUS, TRANSIENT, AND SUBTRANSIENT REACTANCES

The analysis above assumes a steady state operation. However, for example, during and after a short-circuit fault, the rotor oscillations ensue, prior to the rotor coming to another steady state. During these rotor oscillations, the synchronous generator is under a transient condition.

To study transient phenomena, the steady state per-phase equivalent circuit shown in Figure 9.10b needs to be modified. Modeling of synchronous machines can be carried out with increasing levels of complexities, resulting in increasing accuracy. However, in most fault analysis, an estimate of the fault current is needed. Similarly in stability studies, most often all that is desired is to determine if the system would remain stable subsequent to a fault and the time it takes to isolate it. Therefore in these studies, most often a model termed Constant-Flux Model suffices, at least for our educational purposes here. A discussion of this model and the resulting equivalent circuit for use in such studies are presented below.

9.7.1   Constant-Flux Model

The field winding of a synchronous generator is supplied by a DC voltage source V_f such that in steady state, it results in the desired field current $I_f (= V_f/R_f)$ whose DC value is determined by the field resistance R_f. Assuming this DC-excitation voltage to be constant during transients because it cannot change very rapidly, the field winding is essentially a short-circuited coil with a very small resistance, relative to its self-inductance L_{ff}, and thus has a large time-constant. According to the Theorem of Constant Flux-Linkage, the flux linkage of a short-circuited coil remains constant; that is, it cannot change very quickly. Therefore, under brief transient conditions as the stator currents suddenly change causing the armature-reaction flux to change, the field current also suddenly changes to an appropriate value to keep the flux-linkage of the field winding constant.

In steady state, the armature-reaction flux can penetrate the rotor. Therefore, the steady state phasor diagram of Figure 9.10a and the per-phase equivalent circuit of Figure 9.10b were obtained using the principle of superposition. Also, since the armature-reaction flux in steady state can penetrate the rotor, the resulting synchronous reactance X_s is large, generally nearly 1 per unit on the generator base. The field flux-linkage in steady state is the sum of the field flux produced by I_f and the armature-reaction flux going through it. At the armature terminal in steady state, the conditions are $\overline{E}_a$ and $\overline{I}_a$, as shown in Figure 9.10a.

However, immediately following an electrical disturbance such as a short-circuit fault on the electrical network causing the stator currents to suddenly change, the machine is considered in a subtransient condition. In this subtransient condition, the armature reaction flux cannot enter the field winding due to the Constant Flux-Linkage theorem mentioned earlier and much of the armature reaction flux is forced to flow through the air gap, thus resulting in a path of higher magnetic reluctance and hence a smaller reactance. Therefore, the subtransient reactance X_s'' is much smaller than X_s. The ratio between the two can be as large as 4 to 7, where X_s is around 1 per-unit on the machine base. The armature current and the field current, subsequent to a sudden three-phase short circuit at the terminals, are shown in Figure 9.16a and Figure 9.16b, respectively. The subtransient interval corresponds to a few cycles subsequent to the fault, where the armature current is

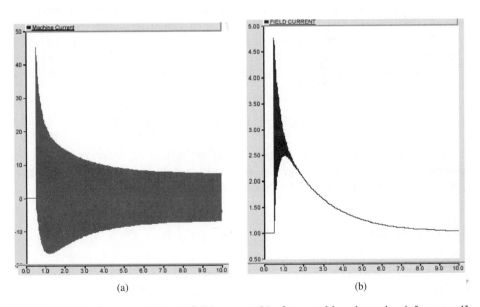

(a)   (b)

FIGURE 9.16 Armature (a) and field current (b) after a sudden short circuit [source: 4].

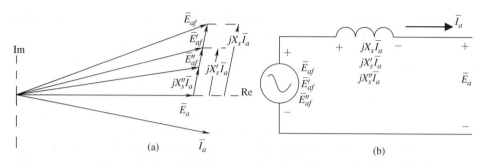

FIGURE 9.17 Synchronous generator modeling for transient and subtransient conditions (in this simplified model, $X_s = X_d$, $X_s' = X_d'$ and $X_s'' = X_d''$).

very large and the field current suddenly jumps to maintain the field-flux linkage constant.

A few cycles after the electrical disturbance causing the stator currents to change but prior to steady state, the machine is considered in a transient state. In this condition, some flux lines manage to penetrate the field winding and the resulting transient reactance X_s' has a value such that $X_s'' < X_s' < X_s$, where X_s' is generally twice of X_s''.

In simplified models, the saliency effects can be neglected. Therefore, $X_s = X_d$, where X_d is the direct-axis synchronous reactance. Similarly, $X_s' = X_d'$ and $X_s'' = X_d''$, where X_d' and X_d'' are the direct-axis transient and the direct-axis subtransient reactances, respectively.

To make use of the subtransient and the transient conditions in fault-current calculations and transient stability studies respectively, we will modify the steady state per-phase equivalent circuit in Figure 9.10b for later chapters. We should note that prior to the fault, the machine is in steady state and the terminal conditions $\bar{E}_a$ and $\bar{I}_a$ are as shown in Figure 9.10a, resulting in the per-phase equivalent circuit of Figure 9.10b. Therefore, our model should be such that it results in the appropriate voltage and current at the machine terminal in the steady state, and yet it is valid in the subtransient and the transient conditions. This can be accomplished in Figure 9.10 by modifying the phasor diagram and the per-phase equivalent circuit as shown in Figure 9.17, where

$$\bar{E}_{af} = \bar{E}_a + jX_s\bar{I}_a \qquad \bar{E}_{af}' = \bar{E}_a + jX_s'\bar{I}_a \qquad \bar{E}_{af}'' = \bar{E}_a + jX_s''\bar{I}_a \qquad (9.13)$$

Therefore, for a given terminal voltage $\bar{E}_a$ and the current $\bar{I}_a$, ignoring the machine resistance, the field-induced voltage can be calculated from Equation 9.13 by using the appropriate reactance based on the type of condition being studied.

REFERENCES

1. N. Mohan, *Electric Machines and Drives: A First Course*, John Wiley & Sons, 2011.
2. P. Anderson, *Analysis of Faulted Power Systems*, Wiley-IEEE Press, 1995.
3. Prabha Kundur, *Power System Stability and Control*, McGraw-Hill, 1994.
4. PSCAD/EMTDC (https://pscad.com/index.cfm?).

PROBLEMS

9.1 Assume that the rotor position in Figure 9.8a corresponds to time $t = 0$. Plot e_{af}, e_{bf}, and e_{cf} as functions of $\omega_{syn}t$.

9.2 Assume that the rotor position in Figure 9.8a corresponds to time $t = 0$. Draw the rotor flux-density space vector $\vec{B}_f$ at $\omega_{syn}t$ equal to 0, $\pi/6$, $\pi/3$, and $\pi/2$ radians.

9.3 If the stator phase current angle θ in Equation 9.4 is zero, then plot the armature-reaction flux-density space vector $\vec{B}_{AR}$ at $\omega_{syn}t$ equal to 0, $\pi/6$, $\pi/3$, and $\pi/2$ radians.

9.4 In the problem of 9.3, plot i_a and $e_{a,AR}$ as functions of $\omega_{syn}t$.

9.5 In the per-phase equivalent circuit of Figure 9.10b, assume $R_s = 0$ and $X_s = 1.2$ pu. The terminal voltage $\overline{V}_a = 1\angle 0$ pu and $\overline{I}_a = 1\angle -\pi/6$ pu. Calculate $\overline{E}_{af}$ and draw a phasor diagram similar to Figure 9.10a.

9.6 In Problem 9.5, with E_{af} kept constant at the magnitude calculated and $V_a = 1$ pu, calculate the maximum power in per unit that this machine can supply.

9.7 In a synchronous generator, assume $R_s = 0$ and $X_s = 1.2$ pu. The terminal voltage $\overline{V}_a = 1\angle 0$ pu. It is supplying 1 pu power. Calculate all the relevant quantities to draw the phasor diagrams in Figure 9.13 if the synchronous generator field-excitation is controlled such that the reactive power Q is as follows: (a) $Q = 0$, (b) supplying $Q = 0.5$ pu, and (c) absorbing $Q = 0.5$ pu.

9.8 Repeat Problem 9.7, assuming that the synchronous machine is a synchronous condenser where the real power $P = 0$.

9.9 In the per-phase equivalent circuit of Figure 9.17b, assume $R_s = 0$, $X_s = 1.2$ pu, $X_s' = 0.33$ pu and $X_s = 0.23$ pu. The terminal voltage $\overline{V}_a = 1\angle 0$ pu and the current $\overline{I}_a = 1\angle -\pi/6$ pu. Draw the phasor diagram similar to Figure 9.17a for the steady state, transient and the subtransient operation.

PSCAD/EMTDC-BASED PROBLEM

9.10 Simulate a sudden short circuit at the terminals of a synchronous generator, as described on the accompanying website.

10

VOLTAGE REGULATION AND STABILITY IN POWER SYSTEMS

10.1 INTRODUCTION

As transmission lines are loaded more towards their capacities, the voltage stability has become a serious consideration. There have been several blackouts caused by voltage collapse. In this chapter, we will examine the causes of voltage stability, the role of reactive power in maintaining voltage stability, and means to supply reactive power.

10.2 RADIAL SYSTEM AS AN EXAMPLE

To understand voltage dependence phenomenon, consider a simple radial system as shown in Figure 10.1a, where an ideal source is supplying a load through a transmission line of series reactance X_L and shunt susceptances as shown. For simplification, the transmission line resistance is ignored. To analyze such a system, the susceptances on both sides are combined as parts of the sending-end and the receiving-end systems and represented by the equivalent circuit in Figure 10.1b. To analyze real and reactive powers in the equivalent system of Figure 10.1b, assuming the receiving-end voltage to be the reference—that is, $\overline{V}_R = V_R \angle 0$:

$$\overline{I} = \frac{\overline{V}_s - V_R}{jX_L} \tag{10.1}$$

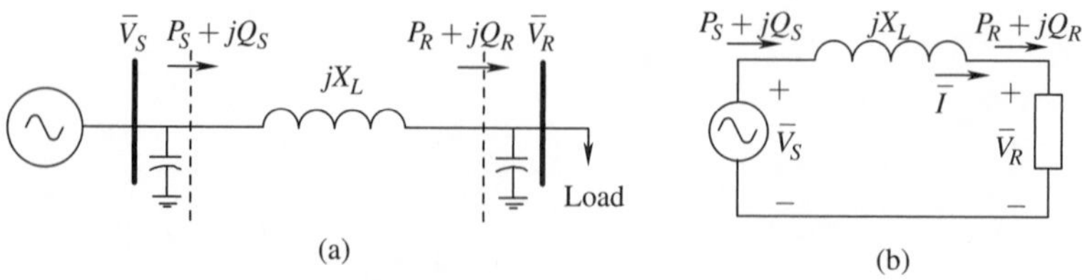

FIGURE 10.1 A radial system.

At the receiving end, the complex power can be written as

$$S_R = P_R + jQ_R = V_R \overline{I}^*$$ (10.2)

Using the complex conjugate from Equation 10.1 into 10.2 and expressing $\overline{V}_S$ in its polar form as $\overline{V}_S = V_S \angle \delta$,

$$P_R + jQ_R = V_R \left(\frac{V_S \angle (-\delta) - V_R}{-jX_L} \right) = \frac{V_S V_R \sin \delta}{X_L} + j \left(\frac{V_S V_R \cos \delta - V_R^2}{X_L} \right)$$ (10.3)

Equating the real parts on both sides of the equation,

$$P_R = \frac{V_S V_R}{X_L} \sin \delta$$ (10.4)

where, assuming no transmission-line losses, P_R is the same as the sending-end power P_S. And

$$Q_R = \frac{V_S V_R \cos \delta}{X_L} - \frac{V_R^2}{X_L}$$ (10.5)

Dividing both sides of Equation 10.5 by $\frac{V_R^2}{X_L}$ and rearranging terms,

$$\frac{V_R}{V_S} = \cos \delta \left(\frac{1}{1 + \frac{Q_R}{V_R^2/X_L}} \right)$$ (10.6)

In power systems, utilities try to keep bus voltage magnitudes close to their nominal values of 1 per unit. Therefore from Equation 10.4, higher values of the transmission-line loading P_R would require higher values of $\sin \delta$, and therefore lower values of $\cos \delta$. Hence, to maintain both voltages close to one per unit, the reactive power Q_R must be negative, that is, at higher power loading of a transmission line, the receiving-end must supply reactive power locally to maintain its bus voltage.

The above requirement for the receiving-end to supply reactive power to maintain its voltage can be further explained by the phasor diagram of Figure 10.2a. By Kirchhoff's Voltage Law in Figure 10.2b,

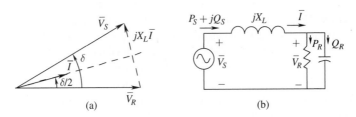

(a) (b)

FIGURE 10.2 Phasor diagram and the equivalent circuit with $V_S = V_R = 1$ pu.

$$\overline{V}_S = \overline{V}_R + jX_L\,\overline{I} \tag{10.7}$$

Assuming both bus voltage magnitudes at 1 per unit and $\overline{V}_R$ as the reference voltage, the phasor diagram is shown in Figure 10.2a where $\overline{V}_S$ leads by an angle δ.

The angle between the two voltages depends on the power transfer over the line, given by Equation 10.4. If both bus voltages are of equal magnitude, $\overline{I}$ from Equation 10.1 is as shown in Figure 10.2a, exactly at an angle $\delta/2$. This phasor diagram clearly shows $\overline{I}$ in Figure 10.2a leading $\overline{V}_R$, implying that Q_R is negative. This means that to achieve V_R equal to 1 per unit voltage, similar to that at the sending-end, the receiving-end should "equivalently" appear as shown in Figure 10.2b, where the equivalent resistance absorbs P_R and the equivalent capacitor supplies the reactive power equal to $|Q_R|$. Higher the load, greater would be δ and I, resulting in a higher demand for the reactive power.

It is useful to know as well what is happening at the sending-end of the transmission line. From the phasor diagram of Figure 10.2a, it can be observed that the sending-end reactive power is the same in magnitude as Q_R but opposite in polarity

$$Q_S = -Q_R \tag{10.8}$$

Therefore, the sending-end supplies reactive power—for example, the generator at the sending-end operates overexcited and the sending-end side susceptance contributes to it as well to a certain extent.

In the transmission line, the reactive power consumed can be calculated as

$$Q_{Line} = I^2 X_L \tag{10.9}$$

Just like real powers, the reactive power supplied to a system must equal the sum of the reactive powers consumed, and thus

$$Q_S = Q_R + \underbrace{I^2 X_L}_{Q_{Line}} \tag{10.10}$$

Using Equation 10.8 into Equation 10.10, the reactive power consumed by the line is twice $|Q_R|$:

$$I^2 X_L = Q_{Line} = 2|Q_R| \tag{10.11}$$

In Figure 10.1, the transmission line is represented by lumped elements and the following discussion above shows what happens at the terminals, with which we are mainly concerned. However, the transmission line has distributed parameters as shown in Figure 10.3a. Assume that the voltages at the two ends are maintained at 1 per unit. If this transmission line, assumed to be lossless, is surge-impedance loaded to $P_R = SIL$, then the voltage profile along the transmission-line length would be flat, as shown by the solid line in Figure 10.3b where the reactive power consumed per unit line length is supplied by its distributed shunt capacitance. Under a heavy load condition with $P_R > SIL$, the voltage profile would sag as shown in Figure 10.3b, and reactive power must be supplied by both ends. The opposite is true under light loadings with $P_R < SIL$, as shown in

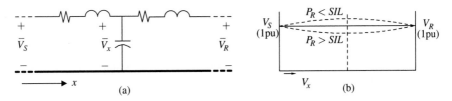

FIGURE 10.3 Voltage profile along the transmission line.

Figure 10.3b, where the reactive power supplied by the transmission-line shunt capacitances must be absorbed at both ends to hold voltages at 1 per unit.

10.3 VOLTAGE COLLAPSE

Once again, let us consider a radial system similar to Figure 10.1b, shown in Figure 10.4a, supplying a load at the receiving end. To begin with, consider a unity power factor load with $Q_R = 0$.

From Equation 10.5,

$$V_R = V_S \cos \delta \tag{10.12}$$

and substituting it into Equation 10.4,

$$P_R = \frac{V_S^2}{X_L} \cos \delta \sin \delta \tag{10.13}$$

To determine the maximum power transfer, taking the partial derivative in Equation 10.13 with respect to angle δ and substituting it to zero results in

$$\frac{\partial P_R}{\partial \delta} = \frac{V_S^2}{X_L} \left(\cos^2 \delta - \sin^2 \delta \right) = 0 \tag{10.14}$$

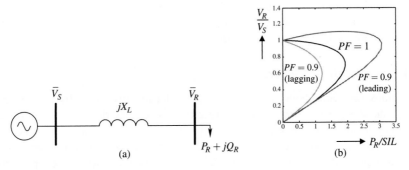

FIGURE 10.4 Voltage collapse in a radial system (example of a 345-kV line, 200 km long).

from which

$$\delta = \pi/4 \tag{10.15}$$

Therefore the maximum power transfer occurs at $\delta = \pi/4$, and using this condition in, Equation 10.13,

$$P_{R,\max} = \frac{V_s^2}{2X_L} \tag{10.16}$$

and

$$V_R \simeq 0.7 \, V_s \tag{10.17}$$

For normalizing P_R, it is divided by the surge impedance loading, SIL. With $Q_R = 0$, corresponding to a unity power factor load, the voltage-ratio (V_R/V_S) is plotted in Figure 10.4b as a function of the normalized real power transfer P_R. Similar curves are plotted for loads at lagging and leading power factor of 0.9 for illustration. These "nose" curves show that as the line loading is increased, the receiving end voltage drops and reaches a "critical" point that depends on the power factor, beyond which further loading (by reducing the load resistance) at the receiving end actually results in lower power, until the voltage at the receiving end collapses.

A lagging power factor loading is worse for the voltage stability compared to the unity power factor loading. From Figure 10.4b, it can be seen that obtaining a near 1 per unit receiving end voltage at a lagging power factor load would require the sending-end voltage to be unacceptably high.

As shown in Figure 10.4b, leading power factor loads result in a higher voltage compared to lagging power factor loads. However, even at leading power factor where one would think that the receiving-end voltage is higher than normal and the voltage stability should be of no concern, a slight increase in power can lead to the critical point and a possible voltage collapse.

10.4   PREVENTION OF VOLTAGE INSTABILITY

As the analysis above shows, the voltage instability is the result of highly loaded systems. Although this has been illustrated using a simple radial system, the same manner of analysis can be applied to a highly integrated system. Voltage instability is associated with the lack of reactive power, and therefore it is necessary to have a reactive power reserve. Several such means are discussed below.

10.4.1   Synchronous Generators

Synchronous generators can supply inductive and capacitive vars by their excitation control, and are the primary source of reactive power. As discussed earlier, heavier loading requires more reactive power support at the receiving end as well as the sending end. However, synchronous generators are limited on how much vars they can supply, as shown in Figure 10.5 by a family of curves corresponding to various pressures of hydrogen used for cooling, at the rated voltage. A positive value of Q signifies reactive power supplied by the generator in the overexcited mode. There are three distinct regions

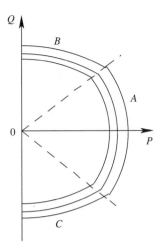

FIGURE 10.5 Reactive power supply capability of synchronous generators.

of reactive power capabilities as shown, as a function of the real power P that depends on the mechanical input.

In region A, the reactive power capability is limited by the heating due to the armature (stator) current and therefore, the magnitude of the apparent power $|S|(= \sqrt{P^2 + Q^2})$ must not exceed its rated value in steady state. In region B, the generator is operating overexcited and is limited by the field-current heating. In region C, the generator is operating underexcited and the end-region heating, as explained in [1], can be a problem that limits the armature current. Generally, the intersection of regions A and B indicates the rating of the synchronous generator in MVA and power factor at the rated voltage. Curves similar to that in Figure 10.5 can be plotted for voltages other than 1 per unit.

In synchronous generators, the conventional excitation control may be too slow to react therefore it is preferable to use a thyristor-based fast-acting excitation control in conjunction with the power system stabilizer (PSS) for damping rotor oscillations. Another possible solution for fast response is to operate the generator overexcited that so normally it is producing more vars than needed, where extra vars are consumed by shunt reactors. Under voltage contingencies, shunt reactors can be quickly disconnected, making those extra vars available to the system.

10.4.2 Static Reactive Power Compensators

Lately for voltage control, power-electronics based static reactive power compensating mechanisms have been proposed and implemented. These are classified under the category of flexible AC transmission systems (FACTS) [4]. Their operating principles are briefly explained in this section.

Need for reactive power in an area can be met by a shunt device as shown in Figure 10.6a, where the system looking into, including the load at that bus where the device is to be connected, can be represented by its Thevenin equivalent, for explanation purposes. The Thevenin impedance is mostly reactive and it is assumed to be purely so, to simplify explanation. From Figure 10.6a,

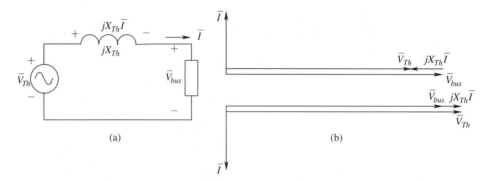

FIGURE 10.6   Effect of leading and lagging currents due to the shunt compensating device.

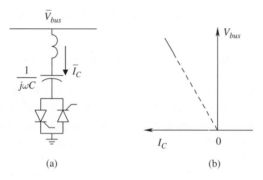

FIGURE 10.7   V-I characteristic of SVC.

$$\overline{V}_{bus} = \overline{V}_{Th} - jX_{Th}\overline{I} \qquad\qquad (10.18)$$

As the phasor diagrams in Figure 10.6b show, if $\overline{I}$ is leading $\overline{V}_{bus}$, then V_{bus} is higher than V_{Th} due to the voltage drop across the Thevenin reactance; the opposite effect occurs if $\overline{I}$ is lagging $\overline{V}_{bus}$. It is important to recognize from Equation 10.18 that the shunt device affects the bus voltage magnitude by means of the drop across X_{Th}: the smaller the Thevenin reactance, the smaller the effect on the bus voltage. For example, if X_{Th} is nearly zero, no amount of current would affect the bus voltage.

The shunt compensating device may consist of capacitor banks which are switched in or out by mechanical means or by back-to-back connected thyristors shown in Figure 10.7a. A small inductance shown in series is mainly to minimize transient current at turn-on. Absence of gate pulses to the thyristors keeps them from conducting, whereas applying continuous gate pulses to both thyristors ensures that the current through the pair can flow in either direction, as if it were a mechanical switch that is switched-on. Often, thyristor-switched capacitor banks are referred as static var compensators (SVC). The $V - I$ characteristic of an SVC is a straight line as shown in Figure 10.7b, where the current magnitude I_C varies linearly with bus voltage as $(\omega C)V_{bus}$.

The shunt compensating device may consist of a shunt reactor as shown in Figure 10.8a. Here, the thyristor pair may be designed to act as a switch, as discussed earlier with shunt

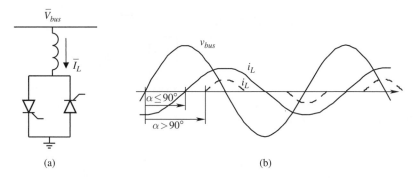

FIGURE 10.8 Thyristor-Controlled Reactor (TCR).

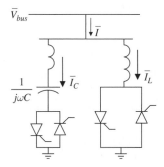

FIGURE 10.9 Parallel combination of SVC and TCR.

capacitor banks, switching the reactor in or disconnecting it. However, it is possible to control each thyristor firing angle, as shown by Figure 10.8b, and hence control the current through the reactor, thus controlling its effective reactance and the vars provided by it. In this mode, this compensating device is called thyristor controlled reactor (TCR). At high voltages, the reactor is fully "in" with a delay angle of 90° or less. Below a certain threshold voltage, the controller begins to increase the delay angle. When the delay angle reaches 180°, the reactor is completely switched out.

It is possible to have both SVC and TCR in parallel as shown in Figure 10.9. By controlling the delay angle of the TCR, the parallel combination can be controlled to either supply or draw reactive power, and its magnitude can also be controlled.

10.4.2.1 STATCOMs

In addition to SVCs and TCRs, it is possible to employ STATCOMs, which are based on voltage-link converters, as shown in Figure 10.10. As discussed in Chapter 7 dealing with HVDC transmission lines, it is possible to synthesize three-phase sinusoidal voltages from a DC source, for example from the DC voltage across the capacitor. A small inductor is at the AC output of the STATCOM, which is not shown in Figure 10.10, is mainly for filtering purposes and has a very little drop across it at the fundamental frequency.

In a STATCOM, the line-frequency voltage is synthesized at the output of the converter from the DC bus voltage across the capacitor; this DC-bus voltage itself is created and maintained by transferring a small amount of real power from the AC system

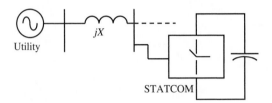

FIGURE 10.10   STATCOM.

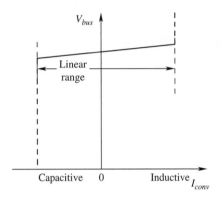

FIGURE 10.11   STATCOM V-I characteristic.

to the converter to overcome converter losses. Otherwise, there is no real transfer of power through the converter, only controlled reactive power as in an inductor or as in a capacitor. A STATCOM can draw controllable capacitive or inductive currents independent of the bus voltage. Therefore, a STATCOM can be considered as a controllable reactive current source on the bus to which it is connected, and its *V-I* characteristic is shown in Figure 10.11, where the vertical lines represent the current rating of the device.

10.4.3   Voltage-Link HVDC Systems

If the power transfer between two systems or areas takes place using a voltage-link HVDC line, then the converters on both sides of a voltage-link HVDC line can independently supply or absorb reactive power, as needed.

10.4.4   Thyristor-Controlled Series Capacitor (TCSC)

Other means for voltage support are series capacitors which reduce the effective value of X_L in the power equation of Equation 10.4. In addition to series capacitors, it is possible to insert a thyristor-controlled series capacitor (TCSC) as shown in Figure 10.12.

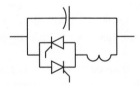

FIGURE 10.12   Thyristor-controlled series capacitors (TCSC) [4].

In TCSCs, the effective inductance of the inductor, in parallel with the capacitor, can be controlled by the conduction angles of the thyristors, and thus controlling the effective reactance of the TCSC to appear as capacitive or inductive. One such unit has been in operation in the western part of the United States [5].

10.4.5 Unified Power Flow Controller (UPFC) and Static Phase Angle Control

In addition to the FACTS devices mentioned earlier, there are additional devices that can directly or indirectly help with the voltage stability. Based on Equation 10.4, a device connected at a bus in a substation, as shown in Figure 10.13a, can influence power flow in three ways by:

1. Controlling the voltage magnitudes
2. Changing the line reactance and/or X
3. Changing the power angle δ

One such device, called the unified power flow controller (UPFC) [4] can affect power flow in any combination of the ways listed above. The block diagram of a UPFC is shown in Figure 10.13a at one side of the transmission line. It consists of two voltage-source switch-mode converters. The first converter injects a voltage $\overline{E}_3$ in series with the phase voltage such that

$$\overline{E}_1 + \overline{E}_3 = \overline{E}_2 \tag{10.19}$$

Therefore, by controlling the magnitude and the phase of the injected voltage $\overline{E}_3$ within the circle shown in Figure 10.13b, the magnitude and the phase of the bus voltage $\overline{E}_2$ can be controlled. If a component of the injected voltage $\overline{E}_3$ is made to be 90 degrees out-of-phase—for example, leading with respect to the current phasor $\overline{I}$, then the transmission line reactance X is partially compensated.

The second converter in a UPFC is needed for the following reason: since converter 1 injects a series voltage $\overline{E}_3$, it delivers real power P_1, and the reactive power Q_1 to the transmission line (where P_1 and Q_1 can be either positive or negative):

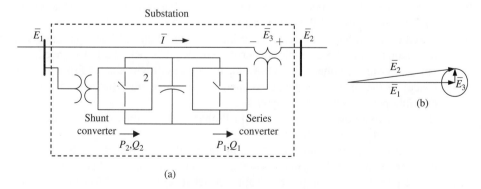

(a)

FIGURE 10.13 Unified Power Flow Controller (UPFC).

$$P_1 = 3\mathrm{Re}\left(\overline{E}_3 \overline{I}^*\right) \tag{10.20}$$

$$Q_1 = 3\mathrm{Im}\left(\overline{E}_3 \overline{I}^*\right) \tag{10.21}$$

Since there is no steady state energy storage capability within the UPFC, the power P_2 into converter 2 must equal P_1 if the losses are ignored:

$$P_2 = P_1 \tag{10.22}$$

However, the reactive power Q_2 bears no relation to Q_1, and can be independently controlled within the voltage and current ratings of the converter 2:

$$Q_2 \neq Q_1 \tag{10.23}$$

By controlling Q_2 to control the magnitude of the bus voltage $\overline{E}_1$, UPFC provides the same functionality as that of an advanced static var compensator STATCOM. A UPFC combines several other functions: Static Var Compensator, Phase-Shifting Transformer and Controlled Series Compensation.

REFERENCES

1. P. Kundur, *Power System Stability and Control*, McGraw-Hill, 1994.
2. C.W. Taylor, *Power System Voltage Stability*, McGraw-Hill, 1994 (for reprints, Email: cwtaylor@ieee.org).
3. *PowerWorld* Computer Program, (http://www.powerworld.com).
4. N. Hingorani, L. Gyugyi, *Understanding FACTS : Concepts and Technology of Flexible AC Transmission Systems*,Wiley-IEEE Press, 1999.
5. W. Breuer, D. Povh, D. Retzmann, Ch. Urbanke, M. Weinhold, "Prospects of Smart Grid Technologies for a Sustainable and Secure Power Supply," The 20th World Energy Congress and Exposition, Rome, Italy, November 11-25, 2007.

PROBLEMS

10.1 In the power flow example of Chapter 5 in the 3-bus power system, what reactive power compensation needs to be provided at bus-3 to bring its voltage to 1 pu.

10.2 In the power flow example of Chapter 5 in the 3-bus power system, what will be the voltage at bus-3 if the power demand at bus-3 is reduced by 50 percent.

10.3 Based on the voltage sensitivity to reactive power change at bus-3, as calculated in Chapter 5 for the example power system, what is reactive power needed at bus-3 to bring its voltage to 1 pu. Compare this result with that of Problem 10.1.

10.4 Calculate the reactive power consumed by the three transmission lines in the example power system of Chapter 5.

10.5 If the reactive power compensation in Problem 10.1 is provided by shunt capacitors, calculate their value.

10.6 If the reactive power compensation in Problem 10.1 is provided by a STATCOM, calculate equivalent $\overline{V}_{conv}$ in pu; assume X as shown in Figure 10.10 to be 0.01 pu.

10.7 In the power flow example of Chapter 5, what will be the effect on the bus-3 voltage if lines 1-3 and 2-3 are 50 percent compensated by series capacitors.

10.8   In the power flow example of Chapter 5, what must be the voltage at bus-1 to bring the bus-3 voltage to 1 pu.

10.9   Why is the performance of STATCOMs superior to that of shunt capacitors?

PSCAD/EMTDC-BASED PROBLEMS

10.10 Model a TCR, as described on the accompanying website.

10.11 Model a TCSC, as described on the accompanying website.

POWERWORLD-BASED PROBLEMS

10.12 Confirm results of Problem 10.1.

10.13 Confirm results of Problem 10.2.

10.14 Confirm results of Problem 10.4.

10.15 Confirm results of Problem 10.5.

10.16 Confirm results of Problem 10.6.

10.17 Confirm results of Problem 10.7.

10.18 Confirm results of Problem 10.8.

11

TRANSIENT AND DYNAMIC STABILITY OF POWER SYSTEMS

11.1 INTRODUCTION

In an interconnected power system such as that in North America, thousands of generators operate normally in synchronism with each other. They share load based on economic dispatch and optimum power flow as discussed in the next chapter. But major disturbances, even momentarily, such as a fault in the system, loss of generation or sudden loss of load can threaten this synchronous operation. Therefore, the ability of the power system to maintain synchronism when subjected to any of the major disturbances previously mentioned is called the transient stability. Such an interconnected system should also have enough damping to retain dynamic stability, as explained later in this chapter.

11.2 PRINCIPLE OF TRANSIENT STABILITY

The principle of transient stability can be illustrated by a simple system with one generator connected through a transformer to an infinite bus, considered to be an ideal voltage source with a voltage of $\overline{V}_B(= V_B \angle 0)$, as shown in Figure 11.1a by two parallel lines.

Power being delivered by the mechanical source is P_m, which equals the generator electrical output P_e in steady state, assuming all the generator losses to be zero. As discussed in Chapter 9, under transient conditions, using the constant flux model, a synchronous generator can be represented by a voltage source of a constant amplitude at the back of the transient reactance X'_d of the generator, as shown in Figure 11.1b, where

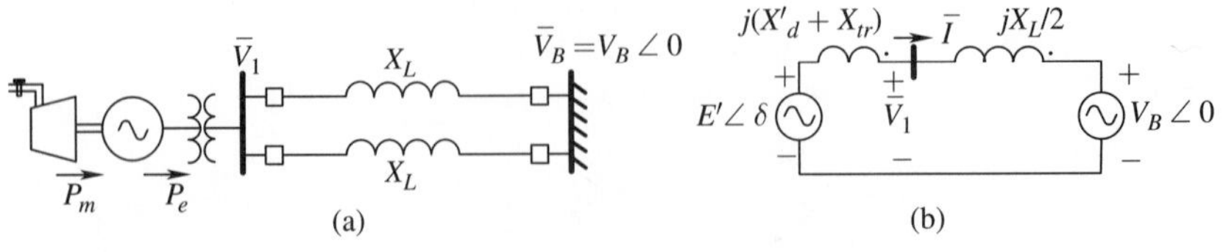

(a)   (b)

FIGURE 11.1   Simple one-generator system connected to an infinite bus.

178

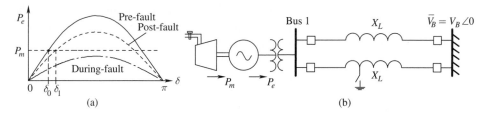

FIGURE 11.2 Power-angle characteristics.

$\overline{E}'(= E' \angle \delta)$ is such that in steady state, prior to the fault, the equivalent circuit of Figure 11.1b results in $P_e = P_m$. In Figure 11.1b, X_{tr} is the leakage reactance of the transformer. Ignoring all the losses in the system of Figure 11.1b, as derived in Chapter 2, the electrical power in MW delivered by the generator to the infinite bus is

$$P_e = \frac{E' V_B}{X_{T1}} \sin \delta \qquad (11.1)$$

where E' and V_B are the magnitudes of the two voltages in kV that are at an angle δ (in electrical radians) apart and are connected through the total reactance X_{T1} (in Ω), which is the sum of the generator transient reactance X_d', the transformer leakage reactance X_{tr}, and the reactance of the two transmission lines in parallel, $X_L/2$. If this transient stability study lasts may be a second or less, it is reasonable to assume as a first-order approximation that the exciter system of the generator cannot respond in such a short time. Therefore the magnitude of the voltage behind the transient reactance can be assumed constant. Similarly, the turbine governor control, be it a steam or a hydro turbine, cannot react within this short time. Hence, the mechanical power input P_m from the turbine to the generator can be assumed to be constant.

In steady state, the generator is rotating at a mechanical speed ω_m (in mechanical radians/s) which equals the synchronous speed. The electrical power output of the generator, P_e, based on Equation 11.1 is plotted in Figure 11.2a as a function of the rotor angle δ (in electrical radians) associated with the generator voltage. In steady state prior to the disturbance, $P_e = P_m$ and the initial rotor angle is δ_0, as shown in Figure 11.2a.

If a fault were to occur, for example on one of the transmission lines, as shown in Figure 11.2b, then during the fault, the capability to transfer electrical power transfer P_e, shown by the dashed-dotted curve in Figure 11.2a, goes down because the bus-1 voltage goes down. It takes a finite clearing time to isolate the faulted line by the circuit breakers at both ends, and subsequent to the fault-clearing time, the generator and the infinite bus are connected by the remaining transmission line. Using an equation similar to Equation 11.1, the power-angle curve for the post-fault curve is shown dotted in Figure 11.2a. It is clear from Figure 11.2a that if the system is stable following the faulted line taken out of service, the new steady state rotor-angle value will be δ_1. In this section, we will look at the dynamics of how the rotor-angle reaches this new steady state, assuming that the transient stability is maintained.

11.2.1 Rotor-Angle Swing

During and following a disturbance, until a new steady state is reached, $P_e \neq P_m$. As explained in Appendix 11A, this will result in the electrical torque T_e that does not equal the mechanical torque T_m, and therefore the difference of these torques will cause the rotor speed ω_m to deviate slightly from the synchronous speed. Therefore, the rotor angle δ_m in mechanical radians per second can be described as

$$J_m \frac{d^2\delta_m}{dt^2} = T_m - T_e \tag{11.2}$$

where J_m is the moment-of-inertia of the rotational system, which is acted upon the acceleration torque, which is the difference of the mechanical torque input T_m and the electrical generator torque T_e opposing it. As described in Chapter 9, we should note that the rotor-angle δ_m is a measure of the angular displacement or position of the rotor with respect to a synchronously rotating reference axis. Multiplying both sides of Equation 11.2 by the rotor mechanical speed ω_m, and recognizing that the product of the torque and speed equals power, Equation 11.2 can be written as

$$\omega_m J_m \frac{d^2\delta_m}{dt^2} = P_m - P_e \tag{11.3}$$

To express the above equation in per unit, a new inertia-related parameter H_{gen} is defined whose value lies in a narrow range of $3-11$ s for turbo-alternators and in a $1-2$ s range for hydro generators. It is defined as the ratio of the kinetic energy of the rotating mass at the rated synchronous speed $\omega_{syn,m}$ in mechanical radians/s to the three-phase volt-ampere rating $S_{rated,gen}$ of the generator:

$$H_{gen} = \frac{\frac{1}{2}J_m\omega^2_{syn,m}}{S_{rated,gen}} \tag{11.4}$$

Substituting for J_m in terms of H_{gen} defined in Equation 11.4, Equation 11.3 can be written as

$$\left(\frac{\omega_m}{\omega^2_{syn,m}}\right)2H_{gen}\frac{d^2\delta_m}{dt^2} = P_{m,gen,pu} - P_{e,gen,pu} \tag{11.5}$$

where P_m and P_e are in per unit of the generator MVA base $S_{rated,gen}$. Often, the system MVA base S_{system} is chosen to be 100 MVA. Regardless of the chosen value of S_{system}, in per unit of the system MVA base, Equation 11.5 can be written as

$$\left(\frac{\omega_m}{\omega^2_{syn,m}}\right)2H\frac{d^2\delta_m}{dt^2} = P_{m,pu} - P_{e,pu} \tag{11.6}$$

where $P_{m,pu}$ and $P_{e,pu}$ are in per unit of the system MVA base S_{system} and

$$H = H_{gen}\left(\frac{S_{rated,gen}}{S_{system}}\right) \tag{11.7}$$

We should note that even under the transient condition, it is reasonable to assume in Equation 11.6 that the rotor mechanical speed ω_m is approximately equal to the synchronous speed corresponding to the frequency of the infinite bus, that is $\omega_m \approx \omega_{syn,m}$. Therefore, Equation 11.6 can be written as

$$\frac{2H}{\omega_{syn,m}} \frac{d^2\delta_m}{dt^2} = P_{m,pu} - P_{e,pu} \tag{11.8}$$

In Equation 11.8, both the angle deviation and the synchronous speed can be expressed in terms of electrical radians

$$\frac{2H}{\omega_{syn}} \frac{d^2\delta}{dt^2} = P_{m,pu} - P_{e,pu} \tag{11.9}$$

The above equation is called the swing equation that describes how the angle δ swings or oscillates due to unbalance between the mechanical power input and the electrical power output of the generator. To calculate the dynamics of speed and angle as functions of time, several sophisticated numerical methods can be used. However to illustrate the basic principle, we will use the Euler's method, which is the simplest, where we will assume that the integration with time is carried out with a sufficiently small time increment Δt during which $(P_m - P_e)$ can be assumed to be constant. With this assumption, integrating both sides of Equation 11.9 with time,

$$\frac{d\delta}{dt}\bigg|_t = \omega(t) = \omega(t - \Delta t) + \frac{\omega_{syn}}{2H}(P_{m,pu} - P_{e,pu})\Delta t \tag{11.10}$$

Similarly assuming the speed in Equation 11.10 to be constant at $\omega(t - \Delta t)$ during Δt,

$$\delta(t) = \delta(t - \Delta t) + \omega(t - \Delta t)\Delta t \tag{11.11}$$

Equations 11.10 and 11.11 show the rotor dynamics in terms of its speed ω and the angle δ as functions of time. In a simple system with one generator connected to an infinite bus, Example 11.1 below shows the rotor dynamics.

Example 11.1
Consider a simple system discussed earlier in Figure 11.2b. The infinite-bus voltage is $\overline{V}_B = 1\angle 0$ pu. The bus 1 voltage magnitude is $V_1 = 1.05$ pu. The generator has a transient reactance $X'_d = 0.28$ pu at a base of 22 kV (L-L) and its three-phase 1,500 MVA base. On the generator base, $H_{gen} = 3.5$ s. The transformer steps up 22 kV to 345 kV, and has a leakage reactance of $X_{tr} = 0.2$ pu at its base of 1,500 MVA. The two 345-kV transmission lines are 100 km in length, and each has a series reactance of 0.367 Ω/km, where the series resistance and the shunt capacitances are neglected. Initially, the three-phase power flow from the generator to the infinite-bus is 1500 MW.

A three-phase to ground fault occurs on one of the lines, 20 percent of the distance away from bus-1. Calculate the maximum rotor-angle swing δ_m if the fault clearing is 40 ms after which the faulted transmission line is isolated from the system by the circuit breakers at both ends of the line.

Solution This example is solved using a program written in MATLAB which is verified by solving it in PowerWorld [1]. Both of these are described in the accompanying

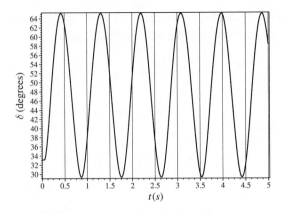

FIGURE 11.3   Rotor oscillation in Example 11.1.

website. The plot of the rotor-angle oscillation is shown in Figure 11.3; these oscillations continue since no damping is included in the model.

11.2.2   Determining Transient Stability Using Equal-Area Criterion

Consider the simple system discussed earlier, repeated in Figure 11.4a where a fault occurs on one of the transmission lines, which is cleared after a time $t_{c\ell}$ by isolating the faulted line from the system. In steady state prior to the fault, $d\delta/dt = 0$ at δ_0, which is the initial steady state value of δ in Figure 11.4b, given by the intersection of the horizontal line representing P_m and the pre-fault curve.

The behavior of the rotor angle can be determined by multiplying both sides of Equation 11.9 by $d\delta/dt$:

$$2\frac{d\delta}{dt}\frac{d^2\delta}{dt^2} = \frac{\omega_{syn}}{H}\left(P_{m,pu} - P_{e,pu}\right)\frac{d\delta}{dt} \qquad (11.12)$$

Replacing δ in the above equation by θ, and integrating both sides with respect to time, we get

$$\int \left(2\frac{d\theta}{dt}\frac{d^2\theta}{dt^2}\right)dt = \frac{\omega_{syn}}{H}\int \left(P_{m,pu} - P_{e,pu}\right)\frac{d\theta}{dt}dt \qquad (11.13)$$

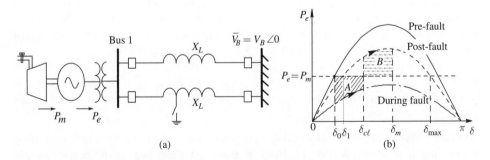

FIGURE 11.4   Fault on one of the transmission lines.

where the integral on the left side equals $(d\theta/dt)^2$. Therefore in Equation 11.13, from the initial angle δ_0 at which $d\delta/dt = 0$, to an arbitrary angle δ

$$\left(\frac{d\delta}{dt}\right)^2 = \frac{\omega_{syn}}{H}\int_{\delta_0}^{\delta}(P_{m,pu} - P_{e,pu})d\delta \tag{11.14}$$

In a stable system, the maximum value of the rotor angle will reach some value δ_m as shown in Figure 11.4b at which instant $d\delta/dt$ once again becomes zero, and then the value of δ begins to decrease. Substituting this condition in Equation 11.14,

$$\frac{\omega_{syn}}{H}\int_{\delta_0}^{\delta_m}(P_{m,pu} - P_{e,pu})d\delta = 0 \tag{11.15}$$

In Equation 11.15 and Figure 11.4b, $P_e = P_{e,\,fault}$ during the fault duration with $\delta_0 < \delta < \delta_{c\ell}$ and $P_e = P_{e,post-fault}$ after the fault is cleared with $\delta_{cl} \leq \delta \leq \delta_m$. Therefore, Equation 11.15 can be written as

$$\underbrace{\int_{\delta_0}^{\delta_c}(P_{m,pu} - P_{e,fault,pu})d\delta}_{Area\ A} - \underbrace{\int_{\delta_c}^{\delta_m}(P_{e,post-fault,pu} - P_{m,pu})d\delta}_{Area\ B} = 0 \tag{11.16}$$

which shows that in a stable system, area A equals area B in magnitude.

During the faulted line still connected to the system, $d\delta/dt$ given by Equation 11.14 is positive in Figure 11.4b, since the mechanical power input exceeds the electrical power output and the rotor angle increases from δ_0 to a new value $\delta_{c\ell}$ at the clearing-time t_{cl}. Area A given by Equation 11.16 and illustrated graphically in Figure 11.4b represents the excess energy delivered to the inertia of the rotor, causing the rotor angle to increase.

After the clearing-time $t_{c\ell}$, at which time the rotor-angle has reached $\delta_{c\ell}$, circuit breakers at both ends of the faulted transmission line open up to isolate the faulted-line from the rest of the system and the electrical output shifts to the post-fault power-angle curve, shown dotted in Figure 11.4b. Now, the electrical power output exceeds the mechanical power input in the second part of the integral in Equation 11.16. Therefore beyond $\delta_{c\ell}$ in Figure 11.4b, the speed begins to decrease although still above the synchronous speed, causing the angle to still keep on increasing, as shown in Figure 11.4b. When area A equals area B in magnitude, the excess energy supplied to the rotor equals that given up by it, and the rotor speed return to its original synchronous speed value. At this time, the rotor angle reaches its maximum value δ_m and at this instant, $d\delta/dt = 0$ in Equation 11.14, and the area B equals area A.

Thus the equal-area criterion given by Equation 11.16 is able to determine the maximum swing of the rotor angle. To maintain stability, δ_m, determined by the equal-area criterion in Equation 11.16, must be less than δ_{max}, as labeled in Figure 11.4b, in order for the generator to remain in synchronism, as further explained in section 11.2.2.1. Beyond the time the rotor angle reaches δ_m, the electrical output in this illustration of Figure 11.4a is still greater than the mechanical input, hence the speed begins to decline and the rotor-angle δ begins to decrease, as shown in Figure 11.5.

With the lack of damping assumed here, the angle would oscillate forever around δ_1 in Figure 11.5, between δ_m and δ_2, with areas C and D equal in magnitude.

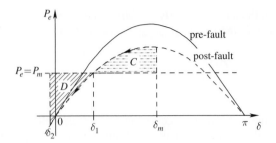

FIGURE 11.5   Rotor oscillations after the fault is cleared.

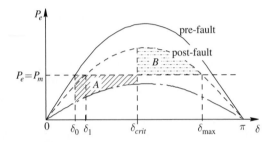

FIGURE 11.6   Critical clearing angle.

However, the damping in a real system would eventually cause the rotor angle to settle down at δ_1.

11.2.2.1   Critical Clearing-Angle

Consider a case as shown in Figure 11.6, where the clearing time is larger than before, such that with areas A and B equal in magnitude, δ_m is equal to δ_{max} labeled in Figure 11.6.

Figure 11.6 represents the limiting case of the critical (maximum) clearing time and hence δ_{crit} at that time, beyond which the stability would be lost. Prior to the rotor angle reaching δ_{max}, if the generator inertia is not able to give up the excess energy acquired during the fault-period, then it will not be able to slow down. The reason is that beyond δ_{max}, the mechanical power input is greater than the electrical power output under the post-fault condition. Therefore, the rotor angle will keep on increasing, resulting in a so-called out-of-step operation. This will cause the relays to trip the circuit breakers to isolate the generator to prevent damage due to excessive currents in the system, and the stability will be lost. Therefore, for a given type and location of a fault or a sudden change in the electrical load, there is a critical angle δ_{crit} corresponding a maximum critical clearing time t_{crit} that results in area A equal to area B in magnitude as shown in Figure 11.6. Note that $\delta_{max} = \pi - \delta_1$.

Of course, the discussion above is theoretical. In practice, there must be a sufficient safety margin to maintain transient stability.

Example 11.2

Consider a simple system discussed earlier in Figure 11.4a. The infinite-bus voltage is $\overline{V}_B = 1 \angle 0$ pu. The voltage magnitude at Bus 1 is $V_1 = 1.05$ pu. The generator has a transient reactance $X'_d = 0.28$ pu at a base of 22 kV (L-L) and 1,500 MVA. On the generator base, $H_{gen} = 3.5\ s$. The transformer steps up 22 kV to 345 kV and has a leakage reactance of $X_{tr} = 0.2$ pu at its own base of 1,500 MVA. The two 345 kV transmission lines are 100 km in length, and each has a series reactance of 0.367 Ω/km, where the series resistance and the shunt capacitances are neglected. Initially, the three-phase power flow from the generator to the infinite-bus is 1500 MW.

A three-phase to ground fault occurs on one of the lines, 20 percent of the distance away from Bus-1. Calculate the maximum rotor-angle swing δ_m if the rotor angle at the time of fault clearing is 50°.

Solution   The solution to this example is carried out by a MATLAB program included on the accompanying website. The power angle curves for the pre-fault, during-fault, and post-fault conditions are as shown in Figure 11.7, where initially $\delta_0 = 33.50°$. The peak values of the power angle curves are calculated as follows on a system MVA base of 100 MVA: $\hat{P}_{e,pre-fault,pu} = 27.17$, $\hat{P}_{e,fault,pu} = 5.78$, and $\hat{P}_{e,post-fault,pu} = 20.51$. Using these values, the power-angle curves are as shown in Figure 11.7.

During the fault, the Area A in Figure 11.7 can be calculated by using Equation 11.16:

$$Area\ A = \int_{\delta_0}^{\delta_{cl}} (P_{m,pu} - \hat{P}_{e,fault,pu} \sin \delta)d\delta$$

$$= P_{m,pu}(\delta_{cl} - \delta_0) + \hat{P}_{e,fault,pu}(\cos \delta_{cl} - \cos \delta_0) \tag{11.17}$$

Similarly, the Area B in Figure 11.7 can be calculated from Equation 11.16:

$$Area\ B = \int_{\delta_{cl}}^{\delta_m} (\hat{P}_{e,post-fault,pu} \sin \delta - P_{m,pu})d\delta$$

$$= \hat{P}_{e,post-fault,pu}(\cos \delta_{cl} - \cos \delta_m) - P_{m,pu}(\delta_m - \delta_{cl}) \tag{11.18}$$

By applying equal-area criterion, the maximum rotor-angle swing $\delta_m = 95.27°$, as shown in Figure 11.7.

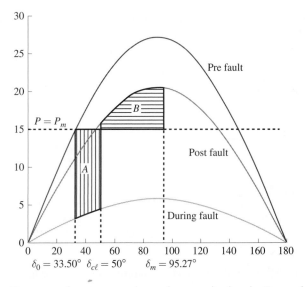

FIGURE 11.7   Power angle curves and equal-area criterion in Example 11.2.

11.3   TRANSIENT STABILITY EVALUATION IN LARGE SYSTEMS

The equal-area method describes the principle behind the transient stability. However in practice, transient stability must be evaluated in the presence of a large number of generators. In such a system, as shown in Figure 11.8 by means of a block-diagram, the rotor electromechanical dynamics is represented in the time-domain and the electrical network is represented in the phasor-domain.

The role of the electromechanical dynamics is to provide the rotor angles (angle of synchronous generator voltages) where as the role of the phasor calculations of the network is to calculate the electrical power that various generators are providing at that time. Phasor calculations are made assuming that the network is in quasi-steady state. This way, the line frequency in the system is eliminated. This procedure, as illustrated in the block diagram of Figure 11.8 allows large time steps to be taken based on the electromechanical time constants of the system.

It is important that the entire network not be calculated in time-domain, which would otherwise require a small simulation time-step and the execution time will be prohibitively large. A time-domain procedure needs to be adopted using a program such as EMTDC only if certain phenomenon such as the performance of thyristor-controlled series-capacitor (TCSC) is being evaluated. The procedure for stability analysis of large networks is illustrated using a three-bus system in Example 11.3.

Example 11.3

Consider a three-bus system discussed earlier in Chapter 5 and repeated in Figure 11.9. These three buses are connected through three 345-kV transmission lines 200 km, 150 km, and 150 km long, as shown in Figure 5.1 of Chapter 5. These transmission lines, considered to consist of bundled conductors, have the line reactance of 0.367 Ω/km at 60 Hz. The line resistance is 0.0367 Ω/km. Ignore all the shunt susceptances. Bus-1 is a slack bus with $V_1 = 1.0$ pu and $\theta_1 = 0$. Bus-2 is a PV bus with $V_2 = 1.05$ pu and $P_2^{sp} = 4.0$ pu. Bus-3 is a PQ bus with the injection of $P_3^{sp} = -5.0$ pu and $Q_3^{sp} = -1.0$ pu.

Both the transformers and the generators have the three-phase MVA ratings of 500 MVA each. Both the generators have a transient reactance $X_d' = 0.23$ pu at a base of

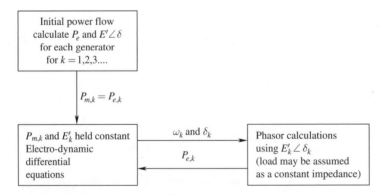

FIGURE 11.8   Block diagram of transient stability program for an *n*-generator case.

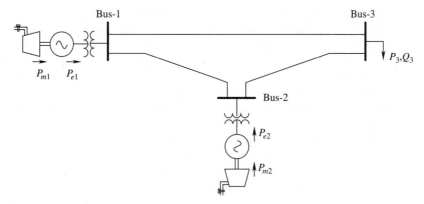

FIGURE 11.9 A 345-kV test example system.

22 kV (L-L) and its own MVA base. Also, each generator has $H_{gen} = 3.5$ s on the generator base. Each 22-kV to 345-kV step-up transformer has a leakage reactance of $X_{tr} = 0.2$ pu at its own MVA base.

A three-phase to ground fault occurs on line 1-2, one-third of the distance away from bus-1. Calculate the rotor-angle swings if the fault-clearing time is 0.1 s after which the faulted transmission line is isolated from the system by the circuit breakers at both ends of the line 1-2.

Solution   The solution to this example using MATLAB and PowerWorld [1] is included on the accompanying website.

11.4   DYNAMIC STABILITY

Without adequate damping, an interconnected power system can develop growing power oscillations [2], causing it to split and possibly resulting in a blackout. One such incident took place in the western USA/Canada system on August 10, 1996, as shown by the plot of power oscillations in Figure 11.10 [3] that left millions of customers without power.

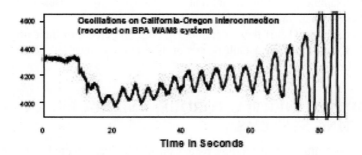

FIGURE 11.10   Growing Power Oscillations: Western USA/Canada system, Aug 10, 1996 [3].

This incident, among many, illustrates that there should be adequate damping in the system so that the oscillations caused by changes in loads, however the size of these changes, decay to a new steady-state operating point. This is considered dynamic stability, which has become particularly important in modern power systems with fast-acting feedback controllers used in excitation systems, HVDC transmission systems and in FACTS (Flexible AC Transmission System [5]) devices. By proper design, for example, using power system stabilizers in conjunction with exciters for synchronous generators and HVDC control [4], it is possible to provide the required damping for maintaining dynamic stability. Series-capacitor compensation of transmission lines can lead to sub-synchronous resonances that can fatigue the turbine-generator shaft, and therefore must be damped out. For the dynamic stability investigation and the design of controllers accordingly, a modal analysis is required that uses the concept of eigenvectors. Therefore, as important as this topic is, it is beyond the scope of the first course on power systems.

REFERENCES

1. PowerWorld Computer Program (www.powerworld.com).
2. Graham Rogers, *Power System Oscillations*, Kluwer, 2000.
3. John F. Hauer et al, Dynamic Performance Validation in the Western Power System, presented at the APEx 2000, Kananaskis, Alberta, Canada, October 2000 (http://certs.lbl.gov/pdf/apex2000.pdf).
4. R.L. Cresap et al, "Operating Experience with Modulation of the Pacific HVDC Intertie," IEEE Transactions on PA&S. Vol. PAS-97, pp. 1053−1059.
5. N. Hingorani, L. Gyugyi, *Understanding FACTS: Concepts and Technology of Flexible AC Transmission Systems*, Wiley-IEEE Press, 1999.

PROBLEMS

11.1   Redo Example 11.1 for a fault clearing time twice as long.

11.2   In Example 11.2, what is the value of the voltage E', as defined in Figure 11.1b.

11.3   In Example 11.2, what is the time taken by the rotor angle to reach $\delta_{cl} = 75°$ at which time the fault is cleared?

11.4   In Example 11.2, what is the value of δ_{crit}, and what is the time taken by the rotor angle to reach that value?

11.5   Redo Example 11.3 if the initial real and reactive powers at Bus-3 are 75 percent of their original values.

11.6   Redo Example 11.3 if the generators H_{gen} are 50 percent of their original value.

11.7   Redo Example 11.3 if a fault occurs on line 1-3, one-third of the distance away from Bus-1.

APPENDIX 11A   INERTIA, TORQUE, AND ACCELERATION IN ROTATING SYSTEMS

Synchronous generators are of rotating type. Consider a lever, pivoted and free to move as shown in Figure 11A.1. When an external force f is applied in a *perpendicular* direction at a radius r from the pivot, then the torque acting on the lever is

$$T_{[Nm]} = f_{[N]} \; r_{[m]} \qquad (11A.1)$$

which acts in a counter clockwise direction, considered here to be positive.

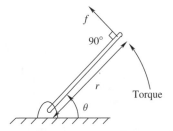

FIGURE 11A.1 Pivoted lever.

In the turbine-generator system, forces shown by arrows in Figure 11A.2 are produced by the turbine. The definition of torque in Equation 11A.1 correctly describes the torque T_m that causes the rotation of the turbine and the synchronous generator connected to it by a shaft.

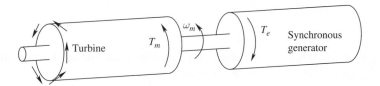

FIGURE 11A.2 Forces and torques in a turbine-generator system.

In a rotational system, the angular acceleration due to a net torque acting on it is determined by its moment-of-inertia J_m of the entire system. The net torque T_a acting on the rotating body of inertia J causes it to accelerate. Similar to systems with linear motion where $f_a = M a$, Newton's Law in rotational systems becomes

$$T_a = J_m \alpha_m \qquad (11A.2)$$

where the angular acceleration $\alpha_m (= d\omega_m/dt)$ in rad/s^2 is

$$\alpha_m = \frac{d\omega_m}{dt} = \frac{T_m}{J_m} \qquad (11A.3)$$

where the damping is neglected. In MKS units, a torque of 1 Nm, acting on an inertia of 1 kg·m^2 results in an angular acceleration of 1 rad/s^2.

In systems such as the one shown in Figure 11A.3a, the turbine produces an electromagnetic torque T_m. The bearing friction and wind resistance (drag) can be combined with the synchronous generator torque T_e opposing the rotation. The net torque, the difference between the mechanical torque developed by the turbine and the generator torque opposing it, causes the combined inertias of the turbine and the generator to accelerate in accordance with Equation 11A.3

$$\frac{d}{dt}\omega_m = \frac{T_a}{J_m}$$ (11A.4)

where the net torque $T_a = T_m - T_e$ is as shown in Figure 11A.3b, and J_m is the equivalent combined inertia.

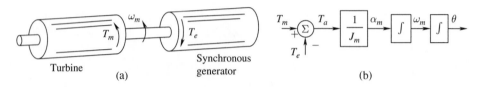

Turbine   (a)   Synchronous generator   (b)

FIGURE 11A.3 Accelerating torque and acceleration.

Equation 11A.4 shows that the net torque is the quantity that causes acceleration, which in turn leads to changes in speed and position. Integrating the acceleration $\alpha(t)$ with respect to time,

$$\text{Speed } \omega_m(t) = \omega_m(0) + \int_0^t \alpha(\tau)d\tau$$ (11A.5)

where $\omega_m(0)$ is the speed at $t=0$ and τ is a variable of integration. Further integrating $\omega_m(t)$ in Equation 11A.5 with respect to time yields

$$\theta(t) = \theta(0) + \int_0^t \omega_m(\tau)d\tau$$ (11A.6)

where $\theta(0)$ is the position at $t = 0$, and τ is again a variable of integration. Equations 11A.4 through 11A.6 indicate that torque is the fundamental variable for controlling speed and position. Equations 11A.4 through 11A.6 can be represented in a block-diagram form, as shown in Figure 11A.3b.

In a rotational system shown in Figure 11A.4, if a net torque T_a causes the cylinder to rotate by a differential angle $d\theta$, the differential work done is

$$dW = Td\theta$$ (11A.7)

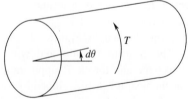

FIGURE 11A.4 Torque, work, and power.

If this differential rotation takes place in a differential time dt, the power can be expressed as

$$p = \frac{dW}{dt} = T\frac{d\theta}{dt} = T\omega_m \qquad (11A.8)$$

where $\omega_m = d\theta/dt$ is the angular speed of rotation.

12

CONTROL OF INTERCONNECTED POWER SYSTEM AND ECONOMIC DISPATCH

12.1 CONTROL OBJECTIVES

In an interconnected power system such as that in North America, thousands of generators, connected through hundreds of thousands of miles of transmission and sub-transmission lines, operate synchronously to supply load that is constantly changing. The main advantage of a highly interconnected system is the continuity of service to consumers, ensuring reliability in case of contingencies such as unscheduled outages. An interconnected system also affords the economy of operation using the optimum generation, making use of the lowest cost generation. As we will see shortly, frequency deviations are also smaller in a highly interconnected system.

In meeting constantly changing load demand and possibly the network structure due to outages, the network voltage and frequency are maintained by the following means:

1. Voltage regulation by excitation control of generators to control the reactive power supplied by them
2. Frequency control and maintaining the interchange of power at their scheduled values
3. Optimal power flow such that the power to the load is provided in the most economic manner, considering the constraints such as the transmission-line capacities and the power system stability

In addition to the controls mentioned above, there are supplementary controls that lead to reducing the integral of the frequency error periodically to zero so that the clocks and other appliances that depend on the grid frequency get back to their normal values and the integral of the power-exchange error is also reduced to zero.

Although the voltage and the frequency controls are implemented simultaneously in time, they act fairly independent of each other as described below.

192

12.2 VOLTAGE CONTROL BY CONTROLLING EXCITATION AND THE REACTIVE POWER

In a power system, the quality of power is defined by keeping the voltage level at its nominal value within a fairly narrow range of ± 5 percent, for example. This is so because the change in the voltage level can disturb the load, at lower voltages dimming lights and slowing down induction motors and at higher voltages causing magnetic saturation of transformers and motors. As discussed in Chapter 10 on voltage stability, lack of reactive power can cause voltage instability and possibly voltage collapse, and to avoid this, certain reactive power reserve must be maintained.

Although there are other means, the primary means of voltage control is by excitation control of synchronous generators in power plants. As discussed earlier, overexciting these generators allows them to supply reactive power to the system, and underexciting them allows them to absorb reactive power. Other means of reactive power control to maintain voltages are the following, which are discussed in the voltage stability chapter: shunt capacitors, shunt inductive reactors, static var controllers (SVCs), STATCOM, series capacitors (including thyristor-controlled series capacitors) and HVDC terminals. Voltages can also be controlled by transformer tap-changing. The last-resort scheme includes automatic load shedding. However, as mentioned earlier, the primary means to obtain voltage control objective is by automatic voltage regulation (AVR) of synchronous generators, as discussed below.

12.2.1 Automatic Voltage Regulation (AVR) through Excitation Control

As discussed in the synchronous generator chapter, the field winding on the rotor is supplied by a DC current to establish the field flux. This requires an excitation system. There are many types of excitation systems in use, but one of the faster-acting excitation systems is shown in Figure 12.1.

It consists of a phase-controlled thyristor rectifier supplied by a three-phase AC derived from the generator output or from an auxiliary supply. The DC output of the thyristor-rectifier is supplied to the field winding, which rotates with the rotor, through brushes and slip rings. A voltage exciter such as the one shown in Figure 12.1 can be designed to regulate the voltage at other than the generator output; for example at the high-voltage bus after the step-up transformer. This can be accomplished by a load

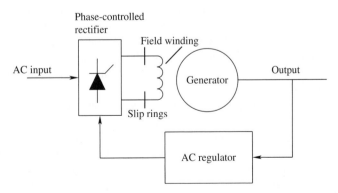

FIGURE 12.1   Field exciter for automatic voltage regulation (AVR).

compensation network which takes account of the voltage drop across the impedance between the generator and the point of regulation, mainly the transformer leakage inductance. In such exciters, several safety features are built in, for example limiting underexcitation to prevent exceeding the steady state stability limit and overexcitation to prevent thermal overloading.

As discussed in the chapter on transient and dynamic stability, a fast excitation response can play a beneficial role in improving transient stability. But, in addition, a power system stabilizer (PSS) should be used to introduce damping to prevent rotor oscillations, by using the rotor-speed oscillations as a signal in controlling the field excitation of the generator to maintain dynamic stability.

12.3 AUTOMATIC GENERATION CONTROL (AGC)

As mentioned earlier, the load demand continuously and randomly fluctuates, and the real power being generated must be adjusted to meet instantaneous change in load demand. For this purpose, an interconnected power system, for example the North American power grid, is divided into four interconnections [2]. To operate safely and reliably, each interconnection comprises of several control areas which continuously monitor and control the power flow. Each area consists of a part, entire, or a group of companies, consisting of many generators. These control areas are interconnected to each other through transmission lines, which are called tie-lines. There is a scheduled interchange of power between these control areas to realize the benefits of having them interconnected. On a dynamic basis, most of the generators from all the areas participate in meeting a change in load demand in any one of the areas. However in steady state, each control area, if it can consistent with the definition of a control area, meets the entire change in load demand within its own area. This requires automatic generation control (AGC) and most of the generators are equipped with this. AGC also requires a certain amount of spinning reserve that can quickly meet the instantaneous change in the load demand that fluctuates randomly.

12.3.1 Load-Frequency Control

Turbines of most generators above a certain rating are governed to provide automatic generation control (AGC). To understand turbine-governor control, consider a single turbine as shown in Figure 12.2a. In such a system, if the electrical load were to increase, then the rotor will slow down, resulting in reduced speed and hence reduced frequency. The frequency decrease is sensed as a feedback signal to the regulator that acts (with a negative sign) on the turbine governor to change the valve position to let more steam in. In steady state, neglecting losses, the mechanical input, the electrical output, and the load power are all the same: $P_m = P_e = P_{Load}$.

By design, the load-frequency plot is a drooping straight line GG, as shown in Figure 12.2b. Initially, at a load equal to P_m, the operating frequency is f_0. If the load were to increase in steady state by an amount ΔP_m, then the frequency decreases by a value equal to Δf which allows, as shown in Figure 12.2a, to let more steam in. (Note that when the frequency decreases, Δf is a negative value and the regulator output is multiplied by a negative sign, thus resulting in a change in power ΔP_m which is positive to satisfy the increase in load.)

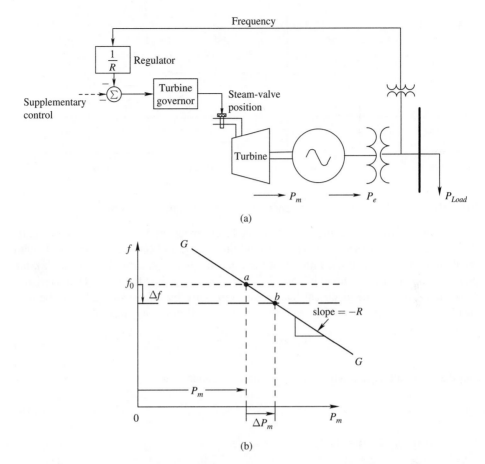

FIGURE 12.2 Load-frequency control (ignore the supplementary control at present).

As shown in Figure 12.2b, the slope of the load-frequency control characteristic is $(-R)$, where R is called the speed regulation, which has a positive value

$$R(\text{in }\%) = -\frac{\Delta f(\text{in }\%)}{\Delta P_m(\text{in pu})} \tag{12.1}$$

or

$$\Delta f(\text{in }\%) = -R(\text{in}\%) \times \Delta P_m(\text{in pu}) \tag{12.2}$$

For example, regulation R equal to 5 percent implies that a 0.1 per unit increase in the electrical load corresponds to 0.5% decrease in the base frequency. If the base frequency is 60 Hz, it corresponds to a decrease of 0.3 Hz, that is, the new frequency will be 59.7 Hz.

Now consider the case of two generators interconnected through a tie-line, supplying the load as shown in Figure 12.3a, where in steady state, neglecting losses, $P_{e1} = P_{m1}$, $P_{e2} = P_{m2}$ and $P_{m1} + P_{m2} = P_{Load}$

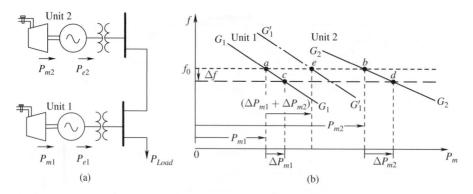

FIGURE 12.3 Response of two generators to load-frequency control.

Both turbines have the governor feedback mechanisms that give their units drooping characteristics as shown by the solid lines G_1G_1 and G_2G_2 in Figure 12.3b, with regulation values R_1 and R_2. Initially at the rated frequency f_0 of 60 Hz (or 50 Hz), the operating points are "a" and "b", and $P_{m1} + P_{m2} = P_{Load}$. For the given characteristics G_1G_1 and G_2G_2, if the electrical load were to increase by an amount ΔP_{Load}, the new steady state frequency (common to both the units) will drop, and from Equation 12.1,

$$\Delta f = -R_1\Delta P_{m1} \quad \text{and} \quad \Delta f = -R_2\Delta P_{m2} \tag{12.3}$$

and the operating points will shift to "c" and "d," such that

$$\Delta P_{m1} + \Delta P_{m2} = \Delta P_{Load} \tag{12.4}$$

From Equations 12.3 and 12.4,

$$\Delta f = -\frac{\Delta P_{Load}}{(1/R_1 + 1/R_2)} \tag{12.5}$$

In general, if many such generators are interconnected, then from Equation 12.5,

$$\Delta f = -\frac{1}{\sum_i 1/R_i}\Delta P_{Load} \tag{12.6}$$

Therefore, comparing Equation 12.6 with Equation 12.2, the equivalent speed regulation is

$$R_{eq} = \frac{1}{\sum_i 1/R_i} \tag{12.7}$$

and the change in frequency for a given change in load is much smaller in an interconnected system, compared to single generator case. As a consequence, interconnected generators result in a much "stiffer" system where all the generators initially participate in accommodating the change in load and thus the change in frequency is very small.

Example 12.1

Consider two generators in parallel operating at 60 Hz and having widely different regulation, with $R_1 = 5\%$ and $R_2 = 16.7\%$. A load change of 0.1 pu occurs. Calculate the equivalent value of the regulation, the initial decrease in frequency, and how the change in load is shared by the two generators initially.

Solution   From Equation 12.7, $R_{eq} = 3.85$ (in %). Therefore, from Equation 12.6,

$$\Delta f = -0.385 \text{ (in \%)} = -0.231 \text{ Hz}$$

After the load change, the new frequency initially will be 59.77 Hz. Using Equation 12.1,

$$\Delta P_{m1} = \frac{0.385}{5.0} = 0.077 \text{ pu} \quad \text{and} \quad \Delta P_{m2} = \frac{0.385}{16.7} = 0.023 \text{ pu}$$

which shows that unit 1 with a smaller value of regulation picks up initially a greater share of the load.

12.3.2 Automatic Generation Control (AGC) and Area Control Error (ACE)

Earlier, we looked at the initial response of the interconnected generator units to a change in load. Now consider two control areas interconnected through a tie-line, as shown in Figure 12.4, where each area may consist of several generator units. In each control area, for discussion purposes, all the generators are combined as an equivalent single generator unit, similar to that in Figure 12.3a.

In order to restore the frequency and the tie-line flow to their original and scheduled values in the system of Figure 12.3a where the load change occurs on unit 1, a supplementary control called the automatic generation control (AGC) raises the generation characteristic of unit 1 in steady state to $G_1'G_1'$ as shown in Figure 12.3b, such that the entire load change is provided by unit 1 and the operating point for unit 1 shifts to a new point "e," whereas the operating point for unit 2 returns to its original value "b."

To do this, for each area, an area control error (ACE) is defined as a sum of the tie-line flow deviation, and the frequency deviation multiplied by a frequency-bias factor B. Therefore, defining the increase in power flow out of one area to the other area—for example the increase ΔP_{12} from area 1 to area 2—to be positive,

$$ACE_1 = \Delta P_{12} + B_1 \Delta f \tag{12.8}$$

Similarly,

$$ACE_2 = \Delta P_{21} + B_2 \Delta f \quad \text{(where neglecting losses, } \Delta P_{21} = -\Delta P_{12}) \tag{12.9}$$

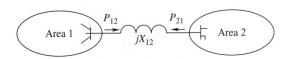

FIGURE 12.4 Two control areas.

A negative value of *ACE* for an area indicates that there is not enough generation in that area. In the block diagram of Figure 12.4, following a load change in either one of the two areas, the final steady state is reached only if both *ACE*, as defined by Equations 12.8 and 12.9, come to zero when the tie-line flow and the frequency are restored to their original values. This implies that in steady state, an area with a load change completely absorbs its own load change. This was earlier illustrated in Figure 12.3b, where the change in load on unit 1 results in shifting the load-frequency characteristic to $G_1'G_1'$ such that it supplies the entire change in that area, and the unit 2 generation remains unchanged at G_2G_2.

Each generator unit participating in the area generation control (AGC) has a supplementary controller that has the area control error (ACE) as the input, as shown in Figure 12.5. The supplementary controller output and the regulation feedback act on the governor to change the steam-valve position, as shown in Figure 12.5.

In steady state, the results are the same regardless of the values of the frequency-bias settings, B_1 and B_2 in Equations 12.8 and 12.9, provided the control is stable. During dynamic operation, field experience has shown that the selection of the frequency-bias factor of approximately $B = 1/R$ in each area provides good dynamic results. When a control area has more than one tie-line, as is almost always the case, the AGC action described above brings the net control-area exchange to its original scheduled value. This is illustrated in the example power system of Figure 12.6, where there are two

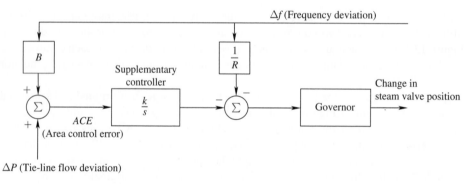

FIGURE 12.5   Area control error (ACE) for automatic generation control (AGC).

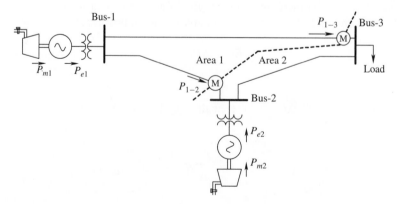

FIGURE 12.6   Two control areas in the example power system with three buses.

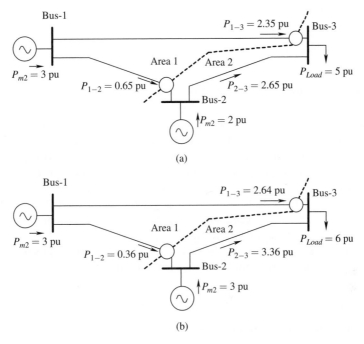

FIGURE 12.7 Line flows in Example 12.2.

control areas connected through two tie-lines. There are two revenue meters (M) to measure power flows at the boundaries of the two control areas. Therefore, in defining ACE in Equations 12.8 and 12.9, the tie-line deviation between the two control areas is $\Delta P_{12} = \Delta P_{1-2} + \Delta P_{1-3}$, where ΔP_{1-2} and ΔP_{1-3} are the line flows at the boundaries of the two control areas.

Example 12.2
Consider the example three-bus power system described in the Power Flow chapter. Ignoring the line losses, calculate the power flow on the three lines. Repeat this when the load has increased to 600 MW (6 pu) but due to AGC applied to both units, the net power flow between the two areas is to remain the same.

Solution Using the MATLAB program developed in the power flow chapter and reducing all line resistances to zero, the line flows are as shown in Figure 12.7a for the load of 500 MW. In case of a 600 MW and both units under AGC, the net flow from Area 1 to Area 2 remains the same, and the entire load change in area 2 is absorbed by unit 2. The line flows are shown in Figure 12.7b.

In practice, the AGC is run every 2 to 4 seconds and the area control error (ACE) has to be brought to zero within every 10-minute interval.

12.3.3 Dynamic Performance of Interconnected Areas

So far, we have examined the interconnections of various areas on a steady state basis. However, the load can change as a step and the interconnected system reacts to it on a dynamic basis. In doing so, it is important that there is enough damping, otherwise, the

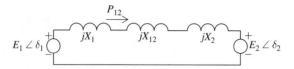

FIGURE 12.8 Electrical equivalent of two area interconnection.

system can go unstable. The dynamic response of the system depends on many parameters: inertias of the interconnected systems, damping, regulators, gain values of the supplementary controller in Figure 12.5 to correct the area control error (ACE), etc.

Another important parameter is the magnitude of the synchronizing torque coefficient between the areas, which can be explained by means of a simple two-area example shown in Figure 12.8 where all the generating units in a control area are represented by an equivalent generator since the objective is to study the inter-area oscillations, not the intra-area oscillations.

In Figure 12.8, X_1 and X_2 are the reactances of each of the equivalent generators with E_1 and E_2 as the internal emfs, and X_{12} is the tie-line reactance. Thus,

$$X_T = X_1 + X_{12} + X_2 \tag{12.10}$$

The power flow from area 1 to area 2 is

$$P_{12} = \frac{E_1 E_2}{X_T} \sin \delta_{12} \tag{12.11}$$

where $\delta_{12} = \delta_1 - \delta_2$. Therefore, linearizing Equation 12.11 around a steady state operating point with initial power as P_0 and the initial angle difference as δ_0, we can write Equation 12.11 as

$$P_0 + \Delta P_{12} = \frac{E_1 E_2}{X_T} \sin(\delta_0 + \Delta \delta_{12}) \tag{12.12a}$$

Noting that $\sin(a + b) = \sin a \cdot \cos b + \cos a \cdot \sin b$, expanding the sine term in Equation 12.12a results in the following for a small value of the perturbation $\Delta \delta_{12}$:

$$P_0 + \Delta P_{12} = \frac{E_1 E_2}{X_T} \left(\sin\delta_0 \cdot \underbrace{\cos\Delta\delta_{12}}_{(\approx 1)} + \cos\delta_0 \cdot \underbrace{\sin\Delta\delta_{12}}_{(\approx \Delta\delta_{12})} \right) \tag{12.12b}$$

Therefore, from Equation 12.12b, recognizing that $\Delta\delta_{12} = \Delta\delta_1 - \Delta\delta_2$,

$$\Delta P_{12} = \underbrace{\left(\frac{E_1 E_2}{X_T} \cos\delta_0 \right)}_{T_{12}} (\Delta\delta_1 - \Delta\delta_2) \tag{12.13}$$

where the quantity within the brackets, if the speed is expressed in per unit as being equal to unity, is the synchronizing torque coefficient T_{12} between areas 1 and 2:

$$T_{12} = \frac{E_1 E_2}{X_T} \cos\delta_0 \tag{12.14}$$

Equation 12.14 shows that the smaller the initial operating angle δ_0, the larger the magnitude of the synchronizing torque coefficient that determines the period and the magnitude of the tie-line oscillations following a load change. These inter-area oscillations in a two-area system are illustrated by an example below.

Example 12.3
Consider a two-area system shown in Figure 12.9 where both areas are identical and both are under AGC. The system parameters are specified in a MATLAB file associated with this example, on the accompanying website. There is a step-change of load in area 1. Plot the following: ΔP_{m1}, ΔP_{m2}, and ΔP_{12}.

Solution   This system is modeled in *Simulink* and included in the accompanying website. The results, which are plotted in Figure 12.10, show that for a load change in area 1, initially both areas participate, but in steady state the load change is entirely satisfied by area 1, and hence $\Delta P_{m1} = \Delta P_{Load1}$ and $\Delta P_{m2} = 0$. The tie-line power deviation ΔP_{12} oscillates and eventually settles down to zero. The frequency of these oscillations depends on the system parameters.

12.4   ECONOMIC DISPATCH AND OPTIMUM POWER FLOW

One of the advantages of an interconnected system is the optimum allocation of generation for the least overall production cost, making sure that the transmission line loadings are within their capacities and the system transient stability margins are maintained.

12.4.1   Economic Dispatch

In power plants, the cost of generating electricity depends on the fixed operating costs (that depend on the capital investment, etc., and are independent of the power being produced), and on the variable operating costs, including fuel costs, which depend on the power being produced. However, once the power plants have been built, the operating strategy from the economic point of view is to minimize the total fuel cost to generate the required amount of power.

In the normal operating range, the efficiency of a thermal power plant increases somewhat with increasing power level. In other words, the heat rate, which is the primary energy in MBTUs (million British thermal units) being consumed per hour divided by the electrical power P decreases slightly with the power level, as shown in Figure 12.11.

Contrary to what the heat-rate curve of Figure 12.11 would suggest, in practice due to various considerations, the fuel cost is taken to increase with power as shown in Figure 12.12a.

In general, the fuel-cost curve for a unit i can be expressed as a quadratic function of the power P_i that is generated:

$$C_i(P_i) = a_i + b_i P_i + c_i P_i^2 \tag{12.15}$$

The slope of the fuel-cost curve in Figure 12.12a at any power level is the marginal cost, which is the cost of generating an additional 1 MWh at a particular power level. This

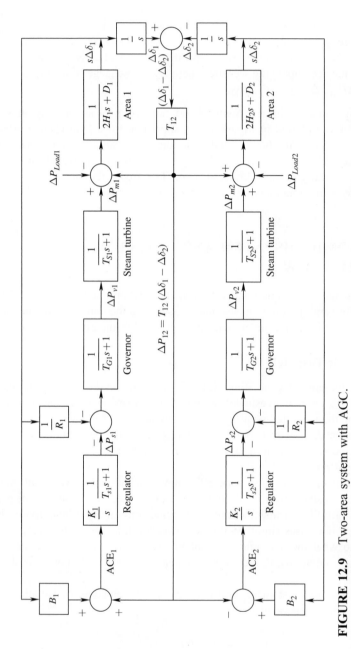

FIGURE 12.9 Two-area system with AGC.
Source: Adapted from Reference 6, Leon K. Kirchmayer, *Economic Control of Interconnected Systems,* John Wiley & Sons, 1959.

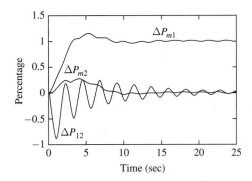

FIGURE 12.10   Simulink results of the two-area system with AGC in Example 12.3.

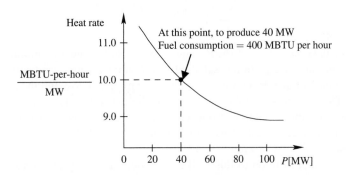

FIGURE 12.11   Heat rate at various generated power levels.

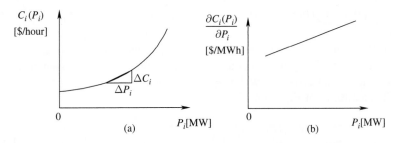

FIGURE 12.12   (a) Fuel cost and (b) marginal cost, as functions of the power output.

marginal cost, in dollars/MWh, can also be calculated by taking the partial derivative of the cost in Equation 12.15 with respect to P_i:

$$\frac{\partial C_i(P_i)}{\partial P_i} = b_i + 2c_i P_i \qquad (12.16)$$

which is a straight line with an upward (positive) slope as shown in Figure 12.12b.

Consider an example where an area has three generators. Then there will be three marginal cost equations similar to Equation 12.16, and there will be three marginal-cost curves, as plotted in Figure 12.13.

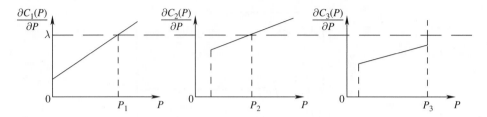

FIGURE 12.13 Marginal costs for the three generators.

Another equation is based on power balance—that is, the sum of the generation must equal the sum of the load and the losses:

$$\sum_i P_i = P_{Load} + P_{Losses} \tag{12.17}$$

We are to calculate the values of P_1, P_2, and P_3 that result in the total minimum fuel cost while the power balance equation is satisfied. An intuitive solution, and correctly so, is to operate such that all three generators have the same marginal cost and the power balance is satisfied. The reason is as follows: If one of the generators is operating at a higher marginal cost, shifting generation from it to the lower marginal-cost generators would result in a lower overall cost. This will continue until all the generators have equal marginal costs, as plotted in Figure 12.13.

A more formal solution to this well-known optimization problem can be obtained using what is called the Lagrangian multiplier method, where a Lagrangian cost function is defined as follows for this three-generator system:

$$L = C_1(P_1) + C_2(P_2) + C_3(P_3) - \lambda[P_1 + P_2 + P_3 - (P_{Load} + P_{Losses})] \tag{12.18}$$

where λ is the Lagrangian multiplier. This Lagrangian function has four variables: P_1, P_2, P_3, and λ, assuming P_{Losses} to be a constant. To minimize this Lagrangian cost function, we will take its partial derivative with respect to each of these four variables and set them to zero. Therefore, from Equation 12.18,

$$\frac{\partial L}{\partial P_1} = \frac{\partial C_1(P_1)}{\partial P_1} - \lambda = 0 \tag{12.19}$$

$$\frac{\partial L}{\partial P_2} = \frac{\partial C_2(P_2)}{\partial P_2} - \lambda = 0 \tag{12.20}$$

$$\frac{\partial L}{\partial P_3} = \frac{\partial C_3(P_3)}{\partial P_3} - \lambda = 0 \tag{12.21}$$

$$P_1 + P_2 + P_3 - (P_{Load} + P_{Losses}) = 0 \quad \left(\because \frac{\partial L}{\partial \lambda} = 0 \right) \tag{12.22}$$

The last equation is the power-balance equation. From 12.19 through 12.22, the cost function in Equation 12.18 is minimized if

$$\frac{\partial C_1(P_1)}{\partial P_1} = \frac{\partial C_2(P_2)}{\partial P_2} = \frac{\partial C_3(P_3)}{\partial P_3} = \lambda \quad \text{(assuming } P_{Losses} \text{ to be constant)} \tag{12.23}$$

The solution given by Equation 12.23 is the same as the intuitively solution: that the overall cost is minimized if all three generators are operating at equal marginal costs while satisfying the power balance. Therefore, from equations similar to Equation 12.16,

$$b_1 + 2c_1 P_1 = \lambda \tag{12.24}$$

$$b_2 + 2c_2 P_2 = \lambda \tag{12.25}$$

$$b_3 + 2c_3 P_3 = \lambda \tag{12.26}$$

and

$$P_1 + P_2 + P_3 = P_{Load} + P_{Losses} \tag{12.27}$$

From these four equations, the four variables P_1, P_2, P_3, and λ can be solved, where λ is the system optimum marginal cost in \$/MWh (that is, the cost in \$/hour of supplying an increase of 1 MW in system load and losses). In practice, there is a maximum limit on the power that can be produced by a generator, and also a lower limit that a unit must produce unless taken offline. These limits can be indicated by vertical lines in the marginal cost curves as shown in Figure 12.13.

Example 12.4

In a control area, there are two generators, 100 MW each. The marginal costs for these two generators can be as expressed as follows:

$$\frac{\partial C_1}{\partial P_1} = 1.8 + 0.01 \, P_1 \quad \text{(in \$/MWh)}$$

$$\frac{\partial C_2}{\partial P_2} = 1.5 + 0.02 \, P_2 \quad \text{(in \$/MWh)}$$

If this area has to supply a total of 150 MW, calculate the area optimum marginal cost λ and the power supplied by each generator.

Solution Equating the two marginal costs and equating the sum of two powers to 150 MW,

$$1.8 + 0.01 \, P_1 = 1.5 + 0.02 \, P_2 \quad \text{and} \quad P_1 + P_2 = 150$$

Solving the above two equations, $P_1 = 90$ MW and $P_2 = 60$ MW. From any of these, the marginal cost is $\lambda = 2.7$ [\$/MWh].

12.4.2 Unit Commitment and Spinning Reserve [4]

The economic dispatch discussed above is based on the generation capacity that has been put into service. However, there are other factors that dictate the capacity that must be in service for a given load, in order for the angle and voltage stability, and for the ability to meet the load change in a control area (now referred to as a balancing authority) quickly. This requires that each control area maintains a certain level, 15 to 20 percent for example, of spinning reserve which represents the aggregate power plant capacity in a control area that is not utilized at a given time, and can respond immediately being on-line, because a cold start, especially of thermal plants, may take a long time—tens of minutes if not hours. If this spinning reserve is not enough, a new generator unit may be committed, and this is called the unit commitment.

12.4.3   Optimal Power Dispatch and Flow

In the analysis above of economic dispatch to minimize overall fuel cost, the transmission line losses are represented as constant, independent of the values of P_1, P_2, and P_3. This is not so, since a generator far away will incur higher transmission line losses and the cost of these losses must be included in the cost minimization problem. Also, no consideration was given to transmission line capacities and the system transient stability. When constraints such as these are factored in, the cost minimization is called optimal power dispatch, and under those conditions, power flow on various lines is the optimal power flow.

REFERENCES

1. Prabha Kundur, *Power System Stability and Control*, McGraw Hill, 1994.
2. Control Area Concepts and Obligations, NERC Document.
3. United States Department of Agriculture, Rural Utilities Service, Design Guide for Rural Substations, RUS BULLETIN 1724E-300.
4. Leon K. Kirchmayer, *Economic Operation of Power Systems*, John Wiley & Sons, 1958.
5. Nathan Cohn, *Control of Generation and Power Flow on Interconnected Systems*, John Wiley & Sons, 1967.
6. Leon K. Kirchmayer, *Economic Control of Interconnected Systems*, John Wiley & Sons, 1959.
7. A. Wood, B. Wollenberg, *Power Generation, Operation, and Control*, 2nd edition, Wiley-Interscience, 1996.

PROBLEMS

12.1   Three generators operating at 60 Hz are connected in parallel and have the following regulation values: $R_1 = 5\%$, $R_2 = 10\%$, and $R_3 = 15\%$. A load change of 0.1 pu occurs. Calculate the equivalent value of the regulation, the initial decrease in frequency, and how the change in load is shared by the three generators initially.

12.2   Repeat Example 12.2 using MATLAB and PowerWorld to compute line flows if the load on bus-3 decreases from 500 MW to 400 MW.

12.3   Repeat Example 12.3 using Simulink, where eliminate the low-pass filtering in the regulators by setting $T_{s1} = T_{s2} = 0$. Compare the results with those in Example 12.3.

12.4   The marginal costs in \$/MWh of three generators can be expressed as follows: $(1.0 + 0.02P)$ for generator 1, $(2.0 + 0.015P)$ for generator 2, and $(1.5 + 0.01P)$ for generator 3, where P is in MW. The total power supplied by these generators is

500 MW. Calculate the system marginal cost λ and the load shared by each of the three generators.

12.5 Repeat Problem 12.4, if the power output of generator 3 is limited in a range of 25 MW to 75 MW.

12.6 The daily load-duration curve over a 24-hour period is as shown in Figure P12.6, where it varies from 200 MW to 600 MW. It is being supplied by three generators whose marginal cost curves are described in Problem 12.4. If an economic dispatch is implemented, calculate the daily fuel cost in dollars for each of the generators and the total, to meet this power demand.

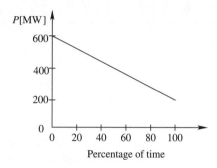

FIGURE P12.6 Load-duration curve.

13

TRANSMISSION LINE FAULTS, RELAYING, AND CIRCUIT BREAKERS

Transmission lines stretch over large distances and are subject to faults involving one or more phases and ground. Such faults cause momentary power outages but, more important, if a protective action is not taken, can cause permanent damage to transmission equipment such as the line itself and the transformers. In this chapter, we will analyze the causes and the types of faults, the relays to detect them and the circuit breakers to isolate these faults and then restore the power system back to normal.

13.1 CAUSES OF TRANSMISSION LINE FAULTS

A common cause of such faults is tree branches near the right-of-way falling on transmission lines and shorting them to ground. As an example, sagging transmission lines touching the trees underneath initiated the major blackout on August 14, 2003, in the northeast United States. Another common fault occurrence is due to backflash when the transmission-line tower or one of the ground wires is struck by lightning that represents a current source of several thousand kilo-amperes. This current flowing through the tower footing impedance can raise the tower potential above the local ground to such a level that without surge arresters (discussed in the next chapter), the insulator strings may flash over.

The lightning current is momentary and lasts just a few tens of micro-seconds, but the arc established due to such an insulator-string flashover results in a short-to-ground of the power-frequency voltages through the arc impedance. If short-circuit currents are not detected by a relay that sends a signal to open circuit breakers to interrupt this current, the current at the power frequency would keep flowing until a serious damage due to fire, for example, would damage the equipment. Therefore, it is essential that these faults be detected and the circuit breakers interrupt them quickly. A few cycles after the fault has been cleared, the circuit breakers can be reclosed and the normal system operation resumes.

The reason to analyze short circuit faults are (1) to set the relays so they can detect it, and (2) to make sure that the circuit breakers ratings are such that they are capable of interrupting the fault currents.

13.2   SYMMETRICAL COMPONENTS FOR FAULT ANALYSIS

More than 80 percent of the faults involve just one of the three phases and ground, for example a tree branch touching or falling on one of the phases. Such faults are often unsymmetrical since all three phases are not involved.

We will begin with the discussion of the analytical approach to analyze unsymmetrical faults of which symmetrical faults are a subset. Consider the fault location f shown in Figure 13.1. To keep this discussion simple yet practical, although the fault may be unsymmetrical, we will assume that the rest of the system "seen" from the fault-location is balanced. Such a fault results in fault currents i_a, i_b, and i_c as shown in Figure 13.1a. During the fault-period, even though the power system voltages and currents are in their transient state, we will assume that they have reached a pseudo-steady state at their line frequency, so that we can make use of phasors to describe them. Therefore, in such a steady state at the line frequency, i_a, i_b, and i_c in Figure 13.1a can be represented by phasors $\bar{I}_a$, $\bar{I}_b$, and $\bar{I}_c$ as shown in Figure 13.1b.

13.2.1   Calculating the Symmetrical Components

To analyze power system with unsymmetrical fault, Fortesque [1] showed long ago that the unbalanced currents can be expressed as sums of components that are symmetrical and balanced. Assuming the power system to be a linear network where the Principle of Superposition can be applied, the voltages and currents in the faulted system can be obtained by summing the components in each of the balanced sequence networks.

As an example, the unbalanced fault currents $\bar{I}_a$, $\bar{I}_b$, and $\bar{I}_c$ are shown in Figure 13.2. They can be shown to be composed of three components as follows:

$$\bar{I}_a = \bar{I}_{a1} + \bar{I}_{a2} + \bar{I}_{a0}$$
$$\bar{I}_b = \bar{I}_{b1} + \bar{I}_{b2} + \bar{I}_{b0} \tag{13.1}$$
$$\bar{I}_c = \bar{I}_{c1} + \bar{I}_{c2} + \bar{I}_{c0}$$

where the subscript 1 refers to the positive-sequence, 2 to the negative-sequence, and 0 to the zero-sequence components.

Within each sequence, the three phases are balanced and sinusoidal at the line frequency. Positive-sequence components $\bar{I}_{a1}$, etc are balanced in amplitude and are of the same a-b-c sequence as the power system voltages and currents, that is phase $\bar{I}_{b1}$ lags $\bar{I}_{a1}$ by 120° and so on. The negative-sequence components are also balanced in amplitude

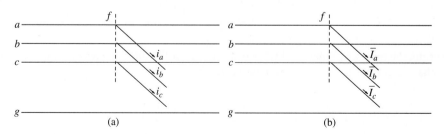

FIGURE 13.1   Fault in power system.

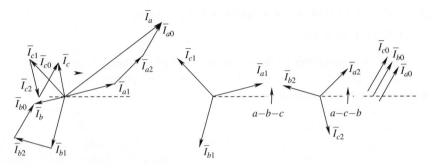

FIGURE 13.2   Sequence components.

but are of the sequence a-c-b, that is, $\bar{I}_{c2}$ lags $\bar{I}_{a2}$ by 120° and so on. The zero-sequence components are also balanced in amplitude, but in contrast to both the positive and the negative-sequence components, $\bar{I}_{a0}$, $\bar{I}_{b0}$, and $\bar{I}_{c0}$ have the same phase, and thus are equal to each other and hence the name zero-sequence. Given these properties of the sequence components, for the ease of writing them, we will define the following operators:

$$a = 1\angle 120° = -0.5 + j0.866$$
$$a^2 = 1\angle 240° = -0.5 - j0.866 \tag{13.2}$$

Therefore, the sequence components in phases b and c in Equation 13.1 can be written in terms of the phase-a components as follows:

$$\bar{I}_{b1} = a^2\bar{I}_{a1}; \quad \bar{I}_{c1} = a\bar{I}_{a1}$$
$$\bar{I}_{b2} = a\bar{I}_{a2}; \quad \bar{I}_{c2} = a^2\bar{I}_{a2} \tag{13.3}$$

Therefore, substituting the above, the fault currents in terms of phase-a components can be written as

$$\bar{I}_a = \bar{I}_{a1} + \bar{I}_{a2} + \bar{I}_{a0}$$
$$\bar{I}_b = a^2\bar{I}_{a1} + a\bar{I}_{a2} + \bar{I}_{a0} \tag{13.4}$$
$$\bar{I}_c = a\bar{I}_{a1} + a^2\bar{I}_{a2} + \bar{I}_{a0}$$

which in a matrix form can be written as

$$\begin{bmatrix} \bar{I}_a \\ \bar{I}_b \\ \bar{I}_c \end{bmatrix} = \begin{bmatrix} 1 & 1 & 1 \\ a^2 & a & 1 \\ a & a^2 & 1 \end{bmatrix} \begin{bmatrix} \bar{I}_{a1} \\ \bar{I}_{a2} \\ \bar{I}_{a0} \end{bmatrix} \tag{13.5}$$

Inverting the matrix in Equation 13.5, we can solve for the three sequence components of the phase-a current in terms of the fault currents:

$$\begin{bmatrix} \bar{I}_{a1} \\ \bar{I}_{a2} \\ \bar{I}_{a0} \end{bmatrix} = \frac{1}{3} \begin{bmatrix} 1 & a & a^2 \\ 1 & a^2 & a \\ 1 & 1 & 1 \end{bmatrix} \begin{bmatrix} \bar{I}_a \\ \bar{I}_b \\ \bar{I}_c \end{bmatrix} \tag{13.6}$$

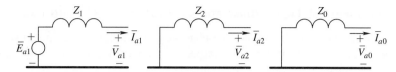

FIGURE 13.3 Sequence networks.

Example 13.1

In Figure 13.2, the three-phase currents are unbalanced and have the following values in per-unit: $\bar{I}_a = 2.2 \angle 26.6°$ A, $\bar{I}_b = 0.6 \angle -156.8°$ A, $\bar{I}_c = 0.47 \angle 138.7°$ A. Calculate the symmetrical components $\bar{I}_{a1}$, $\bar{I}_{a2}$, and $\bar{I}_{a0}$.

Solution Applying Equation 13.6 with the above values of the phase currents, these phase currents are made of symmetrical components with the positive-sequence $\bar{I}_{a1} = 1.0 \angle 15°$ A, the negative-sequence $\bar{I}_{a2} = 0.75 \angle 30°$ A, and the zero-sequence $\bar{I}_{a0} = 0.5 \angle 45°$ A.

13.2.2 Applying the Sequence-Components to the Network and the Superposition

Looking into the system from the fault point for each sequence, the system is balanced and therefore the three-phase system can be represented on a per-phase basis, as discussed in Chapter 3, in terms of phase-a for example. Therefore, for each sequence, the power system network can be drawn on a per-phase basis as shown in Figure 13.3. The connection of these sequence networks depends on the type of the fault. Once the currents and voltages are calculated in these sequence networks for phase-a, they can be used to calculate voltages and currents in the faulted system using equations similar to Equation 13.5.

Note that only the positive-sequence network in Figure 13.3 has an internal emf that equals the Thevenin voltage "seen" from the fault point, looking into the network. The other sequence networks do not have internal emfs because of the assumption of a balanced network prior to the fault.

13.3 TYPES OF FAULTS

There can be many different types of faults and the procedure for solving for all of them is similar. In this chapter, the following faults are considered:

- Symmetrical Three-Phase, and Three-Phase to Ground Fault
- Single-Line to Ground Fault
- Double-Line to Ground Fault
- Double Line Fault (ground is not involved)
- Fault with Fault Impedances

In addition to these short-circuit faults, there can be instances of open-circuit conductor(s) that can be serious safety issues. Analysis of these is left as a homework exercise.

13.3.1   Symmetrical Three-Phase and Three-Phase to Ground Faults

In a balanced system, three-phase or three-phase-to-ground faults are identical where the voltage at the faulted point is zero with respect to ground, and the current flowing into ground is zero since $\bar{I}_a + \bar{I}_b + \bar{I}_c = 0$ in Figure 13.4a.

Since, these three fault currents form a balanced three-phase set, their negative and the zero-sequence components are zero and only the positive-sequence network on a per-phase basis needs to be considered as shown in Figure 13.4b, where, $\bar{I}_a = \bar{I}_{a1}$ since the other two components are zero.

13.3.2   Single-Line to Ground Fault

This is shown in Figure 13.5a, where phase-a is faulted to ground through a fault impedance Z_f.

For this fault,

$$\bar{I}_b = \bar{I}_c = 0 \tag{13.7}$$

$$\bar{V}_a = Z_f \bar{I}_a \tag{13.8}$$

Substituting $\bar{I}_b$ and $\bar{I}_c$ as zero in Equation 13.6,

$$\bar{I}_{a1} = \bar{I}_{a2} = \bar{I}_{a0} \tag{13.9}$$

And thus, from Equation 13.4,

$$\bar{I}_{a1} = \frac{\bar{I}_a}{3} \tag{13.10}$$

From Equation 13.8, in terms of symmetrical components

$$\bar{V}_a = \bar{V}_{a1} + \bar{V}_{a2} + \bar{V}_{a0} = Z_f \bar{I}_a \tag{13.11}$$

Substituting for $\bar{I}_a$ from Equation 13.10 into Equation 13.11,

$$\bar{V}_a = 3Z_f \bar{I}_{a1} \tag{13.12}$$

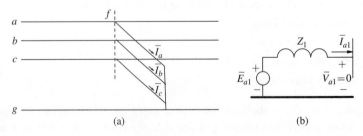

FIGURE 13.4   Three-phase symmetrical fault.

Equations 13.9 and 13.12 are satisfied by connecting the three sequence networks in series as shown in Figure 13.5b, from which

$$\overline{I}_{a1} = \overline{I}_{a2} = \overline{I}_{a0} = \frac{\overline{E}_{a1}}{Z_1 + Z_2 + Z_0 + 3Z_f} \tag{13.13}$$

Knowing the sequence currents, all the three sequence networks can be solved and any of the currents and voltages can be computed in the faulted network.

13.3.3 Double-Line to Ground Fault

This is shown in Figure 13.6a, where phases b and c are faulted to ground. The fault conditions result in the following conditions:

$$\overline{I}_a = 0 \tag{13.14}$$

$$\overline{V}_b = \overline{V}_c = 0 \tag{13.15}$$

Using the condition in Equation 13.15 for voltages into an equation for voltages similar to Equation 13.6,

$$\overline{V}_{a1} = \overline{V}_{a2} = \overline{V}_{ao} \tag{13.16}$$

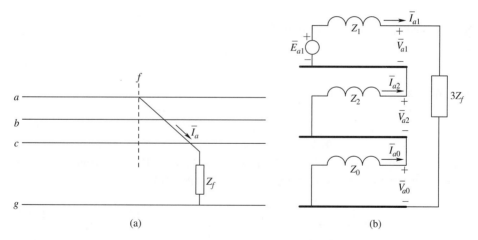

(a)   (b)

FIGURE 13.5   Single-line to ground fault.

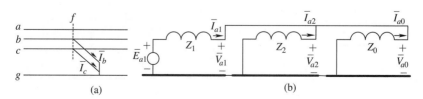

(a)   (b)

FIGURE 13.6   Double line to ground fault.

From Equation 13.16, it is clear all three sequence networks are in parallel as shown in Figure 13.6b, and applying the Kirchhoff's Current Law in this circuit, Equation 13.14 is satisfied in light of Equation 13.4—that is, $\bar{I}_a = \bar{I}_{a1} + \bar{I}_{a2} + \bar{I}_{a0} = 0$.

13.3.4 Double-Line Fault (ground is not involved)

This is shown in Figure 13.7a, where phases b and c are shorted to each other through Z_f. This fault results in the following conditions:

$$\bar{I}_a = 0 \tag{13.17}$$

$$\bar{I}_b = -\bar{I}_c \tag{13.18}$$

$$\bar{V}_b = \bar{V}_c + Z_f \bar{I}_b \tag{13.19}$$

By observation of the faulted system, there is no connection to ground, and hence there cannot be any zero-sequence current

$$\bar{I}_{a0} = 0 \tag{13.20}$$

and hence, in the zero-sequence network of Figure 13.3,

$$\bar{V}_{a0} = 0 \tag{13.21}$$

Since $\bar{I}_a$ and $\bar{I}_{a0}$ are both zero, from Equation 13.4,

$$\bar{I}_{a1} = -\bar{I}_{a2} \tag{13.22}$$

Therefore, from Equation 13.5,

$$\bar{I}_b = (a^2 - a)\bar{I}_{a1} \tag{13.23}$$

From the voltage fault condition of Equation 13.19, with $\bar{V}_{a0} = 0$, from the voltage equation similar to Equation 13.5 for $\bar{V}_b$, using Equation 13.23,

$$\underbrace{a^2 \bar{V}_{a1} + a \bar{V}_{a2}}_{(=\bar{V}_b)} = \underbrace{a \bar{V}_{a1} + a^2 \bar{V}_{a2}}_{(=\bar{V}_c)} + Z_f(a^2 - a)\bar{I}_{a1} \tag{13.24}$$

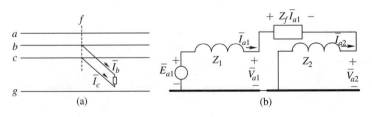

FIGURE 13.7 Double line fault (ground not involved).

From the above equation,

$$\overline{V}_{a1} = \overline{V}_{a2} + Z_f \overline{I}_{a1} \tag{13.25}$$

Using Equations 13.18 and 13.25, the sequence networks are connected as shown in Figure 13.7b.

13.4 SYSTEM IMPEDANCES FOR FAULT CALCULATIONS

For calculating currents under unsymmetrical faults, power system components must be represented by their appropriate impedances for all the three sequences: positive, negative, and zero.

13.4.1 Transmission Lines

Transmission lines are assumed to be perfectly transposed and their positive and negative sequence impedances are the same. Their zero-sequence impedance that involves ground return is greater in value and it can be calculated using a line-constants program such as EMTDC, as described in Chapter 14 dealing with transient overvoltages on transmission lines.

13.4.2 Simplified Synchronous Generator Representation

For calculating fault currents within a few cycles of the fault inception, the sub-transient reactances of the generators, as discussed in Chapter 9, are used. Assuming that it is a round-rotor machine, the positive sequence impedance equals $X_d^{''}$ for the d-axis, whereas that of the q-axis sub-transient reactance may be slightly smaller. Typically, $X_d^{''} = 0.12 - 0.25$ pu.

In the negative sequence, the voltage and currents are of the negative sequence a-c-b, while the rotor is turning at the synchronous speed in the forward direction dictated by the positive sequence a-b-c excitation. In this situation, the armature-reaction mmf produced by the negative-sequence currents would rotate in the direction opposite to the rotor. This would be at twice the synchronous speed with respect to the rotor. Therefore, the d- and q-axis damper windings on the rotor would shield the flux due to the negative-sequence armature-reaction flux from going past the damper windings. Therefore, the negative-sequence reactance of the synchronous generator can be written as

$$X_2 = \frac{X_d^{''} + X_q^{''}}{2} \tag{13.26}$$

Typically, $X_2 \simeq X_d^{''}$.

The zero-sequence impedance depends on the leakage reactance per-phase plus three times the impedance that is usually connected from the neutral-to-ground of generators through a transformer. If the neutral for some reason is floating with respect to ground and this neutral impedance is infinite, then, the zero sequence impedance would also be infinite, indicating that the zero-sequence current cannot flow through the generator. Typically, the internal zero-sequence reactance of a turbo-generator can be approximated as $X_0 = 0.5X_2$.

For fault calculations, usually the positive-sequence network for the generator is represented by the sub-transient reactance and a voltage source behind it, such that together they yield the appropriate pre-fault voltages and currents in the network, as discussed in Chapter 9.

13.4.3   Transformer Representation in Fault Studies

Generally, only the leakage impedance of the transformers needs to be included in fault studies. On a per unit basis, normally the transformer turns-ratio does not appear in calculations. Transformer leakage reactances are the same for the positive and the negative sequences. The zero-sequence impedance depends on how the three phase windings are connected. For example, a delta-connected winding does not provide a path to the zero-sequence currents, as shown in Figure 13.8a.

Similarly, the wye-connected windings with an isolated neutral will appear as an open-circuit to zero sequence, as shown in Figure 13.8b. As shown in Figure 13.8c, a grounded-wye with delta-connected secondary windings provides a short-circuit path that allows zero-sequence currents to flow. This will also be the case if the secondary was connected in a grounded-wye. However, the grounded-wye primary in Figure 13.8c will appear as an open-circuit if the secondary was wye-connected with an isolated neutral. If zero-sequence currents can flow, a neural-to-ground impedance Z_n, as shown in Figure 13.9a, will appear as $3Z_n$ in the zero-sequence network of Figure 13.9b.

In case of three single-phase transformers, the zero-sequence impedance Z_0 in Figure 13.9b equals the positive-sequence leakage impedance. This will also be the case in three-phase transformers with shell-type construction.

Example 13.2
Consider a simple system with a 1 pu load at bus-3 being supplied by a single generator, as shown in Figure 13.10, where all the quantities are in per unit. Calculate the fault

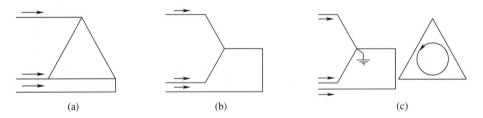

(a)   (b)   (c)

FIGURE 13.8   Path for zero-sequence currents in transformers.

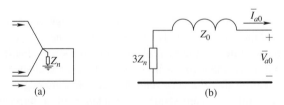

(a)   (b)

FIGURE 13.9   Neutral grounded through an impedance.

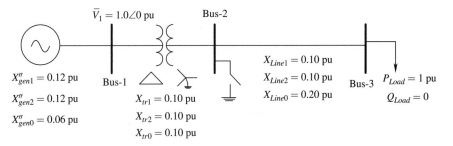

FIGURE 13.10 One-line diagram of a simple power system.

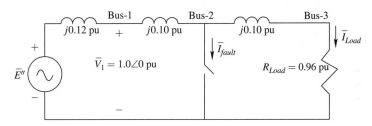

FIGURE 13.11 Positive-sequence circuit for calculating a three-phase fault on bus-2.

current in per unit at bus-2 for (a) three-phase fault, and (b) single-line to ground (SLG) fault with the fault impedance as zero.

Solution Bus-1 is assumed to be the slack bus with $\overline{V}_1 = 1.0 \angle 0$ pu. Using the MATLAB program developed in Chapter 5 on Power Flow or the *PowerWorld*, both of which are included on the accompanying website, the pre-fault voltage at bus-3 is calculated as $\overline{V}_3 = 0.98 \angle -11.79°$ pu. Therefore, the load on bus-3 can be represented by $R_{Load} = 0.96$ pu.

a. In case of a three-phase fault on bus-2, the positive-sequence per-phase circuit is shown in Figure 13.11, where $\overline{I}_{Load}$ and hence $\overline{E}''$ at the back of the sub-transient reactance can be calculated as follows:

$\overline{V}_3 \overline{I}_{Load}^* = 1.0$ pu. Therefore, $\overline{I}_{Load} = 1.02 \angle -11.79°$ pu before the fault. Using $\overline{E}$, with X_{gen1} in series, to represent the generator, and given that $\overline{V}_1 = 1.0 \angle 0°$ pu prior to the fault, we can calculate $\overline{E} = 1.0 \angle 0 + j(0.12 \overline{I}_{Load}) = 1.03 \angle 6.67°$ pu. For a three-phase to ground fault on bus-2, represented by closing the switch in Figure 13.11, $\overline{I}_{fault} = 4.69 \angle -83.32°$ pu.

b. In case of a single-line to ground (SLG) fault on bus-2, the sequence per-phase networks are connected in series as the analysis in Figure 13.5 shows. With the fault impedance $Z_f = 0$, the circuit diagram is shown in Figure 13.12. In the zero-sequence network, the generator zero-sequence impedance is shorted out because of the grounded-wye-delta connection. Also, the 30° phase shift introduced by the transformer connection has no effect on the fault current. From Figure 13.12, $\overline{I}_{fault} = 5.71 \angle -85.7°$ pu. The MATLAB and the PowerWorld files are included on the accompanying website.

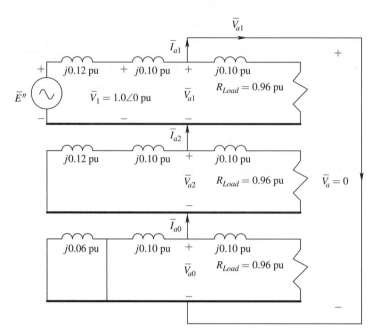

FIGURE 13.12   Sequence networks for calculating the fault current due to SLG fault on bus-2.

13.5   CALCULATION OF FAULT CURRENTS IN LARGE NETWORKS

The above procedure shows the underlying principles for fault calculations which can be used for analyzing a power system with a few buses. In most practical networks with thousands of buses, computer programs have been developed to carry these out. One of the procedures used in computing fault currents in such network is described below.

Earlier, we have discussed the nodal-equation formulation of the network for the positive sequence

$$\overline{I}_{pos} = Y_{pos}\overline{V}_{pos} \qquad\qquad (13.27)$$

where the subscript "pos" is used to indicate positive sequence, in place of "1," which may be a bus number. Equation 13.27 can be written in terms of the impedance matrix as

$$\overline{V}_{pos} = Z_{pos}\overline{I}_{pos} \qquad\qquad (13.28)$$

where $Z_{pos}(= Y_{pos}^{-1})$ is the inverse of the Y-matrix. The impedance matrix in large networks is created without taking the inverse of the Y-matrix. Similar equations can be written to relate the negative-sequence and the zero-sequence voltages and currents. From these networks, the fault currents for any type of fault can be computed using a computer program such as PowerWorld [4].

Example 13.3

Consider the example three-bus power system discussed in earlier chapters and repeated in Figure 13.13. The modeling of this system under the pre-fault operating conditions is described on the accompanying website. A single-line to ground (SLG) fault occurs on the line 1-2, one-third of the distance away from bus-1. Calculate the fault current and various line-currents.

Solution   The solution to this example is calculated by a MATLAB program and the results are verified using the PowerWorld [4] program. Both are included on the accompanying website.

13.6   PROTECTION AGAINST SHORT-CIRCUIT FAULTS

Although, open-circuits sometimes can result in hazardous safety situations to personnel and must be guarded against, it is the overcurrent phenomena due to short-circuit faults is of focus here. To minimize the interval of the power disturbance, and more importantly preventing power equipment from permanent damage, it is important that the fault currents, larger than the load currents for which the apparatus is designed, not be allowed to flow for intervals that can be destructive. For this purpose, all power system apparatus are equipped with overcurrent protection equipment. The entire power system is divided into overlapping zones such that no part of the system is left unprotected. This protection equipment can be categorized as follows, as shown in the block diagram of Figure 13.14 for one of the circuit breakers (CB):

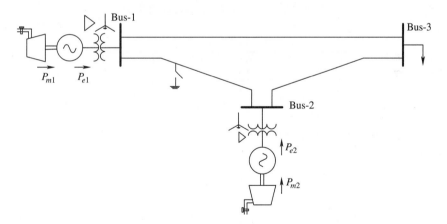

FIGURE 13.13   A SLG fault in the example three-bus power system.

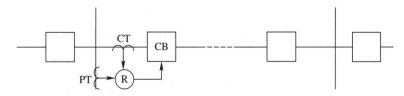

FIGURE 13.14   Protection equipment.

- Current and voltage transformers (CTs and PTs) for sensing these power system voltages and currents
- Relays (R) which determine if a fault has occurred and issue a command to circuit breakers to operate
- Circuit breakers (CB) which open the circuit contacts to interrupt the fault current, and subsequently to reclose them to resume normal operation

All these are described briefly in the subsections below.

13.6.1   Current and Voltage Transformers

Currents and voltages in a power system must be sensed so that relays can determine if a fault has occurred. However, these voltages and currents are at extremely high values and must be stepped down to low-voltage signals with reference to a logic ground.

The current transformer (CT), as shown in Figure 13.15, is a transformer in which the power system current flows through its primary which is usually has a single turn. The secondary generally has a large number of turns and produces a much smaller current, which equals the primary current divided by the turns-ratio, which flows through a small load referred to as "burden." The current-sense voltage signal associated with it is used in the relay logic.

The capacitor-coupled voltage transformer (CCVT), one out of several types, is shown in Figure 13.16. It uses the capacitive voltage-divider principle where the output voltage of this divider is isolated through a transformer for safety purposes.

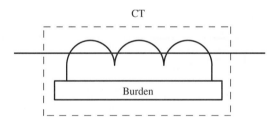

FIGURE 13.15   Current transformer (CT).

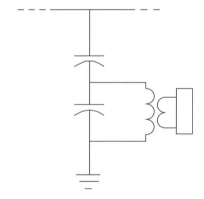

FIGURE 13.16   Capacitor-coupled voltage transformer (CCVT).

13.6.2 Relays

Relays, based on the sensed voltages and currents, other signals like time on, and so on, decide if a fault has occurred and if it should be interrupted by the circuit breaker. It is important that relays operate when they should, in order to protect a power system, but it is equally important that they don't operate falsely in order to avoid causing unnecessary power disturbances. Therefore, in relays, three things are important: selectivity, speed and reliability.

13.6.2.1 Relay Types

Relays are of many types but they can be basically categorized in the following manner:

- *Differential Relays*: These relays may be used, for example to protect a generator, a bus or a transformer against internal faults, as shown in Figure 13.17 for protecting a bus.

  Under normal conditions, as shown in Figure 13.17, the differential current through the relay, which is the difference between the measured currents, is zero. This is not so under an internal fault condition, causing the fault current to trip the circuit breaker.

- *Overcurrent Relays*: In these relays, if the current being measured exceeds a minimum value that is greater than the maximum load current by a certain factor, the relay determines that a fault has occurred, giving a "trip" command to the circuit breaker to operate. Generally, such relays operate on a delay-time basis where this delay time is an inverse function of the magnitude of the fault current; the larger the current magnitude, the shorter the delay time, as shown in Figure 13.18. It is possible to have several settings (1, 2, 3, and so on) as shown in Figure 13.18, thus for the same current,

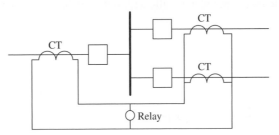

FIGURE 13.17   Differential relay for bus-protection.

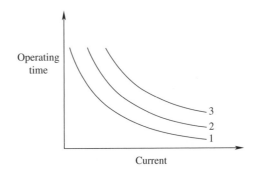

FIGURE 13.18   Time-current characteristics of overcurrent relays.

the relay would operate with different time delays. These settings allow such relays to be co-ordinated with other relays protecting the system.

- *Directional Overcurrent Relays*: The protection offered by this relay is for faults only in one direction. For example in Figure 13.19, if the fault occurs on the right side of the CT location, the current $\bar{I}$ sensed by this relay would be lagging with respect to the voltage at this bus, causing the relay to trip the circuit breaker. Whereas for the fault left of its CT location, the current would be leading with respect to the voltage and the relay would be blocked from tripping the circuit breaker.

- *Ground Directional Overcurrent Relays*: These are zero-sequence relays which, as Figure 13.20 shows for the relay at bus-A, will act instantaneously by issuing a trip command to the circuit breaker if the fault is within 80 percent of the line-section A-B (considered its zone-1 of protection, discussed later); otherwise the tripping time is increased as shown. The overcurrent relay at bus-A is co-ordinated (time-delayed) with adjacent line-section relays. This will ensure, for example, that the relay at bus-A will not trip for faults in the adjacent line-section B-C; it will trip only as a backup to the relay at bus-B, looking into the line-section B-C, if that relay malfunctions and fails to trip. We should note that for a fault on the remaining 20 percent of the line-section A-B, the relay at bus-A is also time-delayed as shown in Figure 13.20; for such a fault, the relay at bus-B, looking towards bus-A, will give an instantaneous trip command to its circuit breaker because this fault will be in its zone-1 of protection, discussed later.

- *Directional Distance (Impedance) Relays*: By calculating the ratio of the measured voltage and current, as shown in Figure 13.19, these directional relays determine the impedance, and hence these are called impedance relays. The characteristic of such relays is plotted in an $R - X$ plane, as shown in Figure 3-21, and it goes through the origin. If the calculated impedance falls within the circle, the relay will trip; otherwise it is blocked.

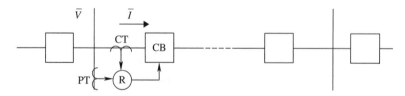

FIGURE 13.19   Directional overcurrent relay.

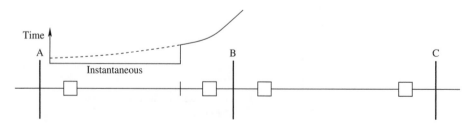

FIGURE 13.20   Ground directional overcurrent relay.

FIGURE 13.21 Directional impedance (distance) relay having a "mho" characteristic.

Under normal system conditions, load currents are much smaller than the fault currents and hence the impedance (ratio of the measured voltage and current) would be much larger, and outside the circle in Figure 13.21 and the relay is blocked from tripping. In case of a fault, for example on the right of the CT location on the line in Figure 13.19, the measured impedance would be the impedance of the line between the relay location and the fault; it would be small, for example somewhere along the straight line in Figure 13.21, and within the circle, causing the relay to trip. This impedance is also an indication of the distance along the line, where the fault has occurred.

The directionality to this relay is given by shifting the characteristic such that it passes through the origin, as shown in Figure 13.21. If the fault occurs to the left of the CT in Figure 13.19, for example, the impedance based on the measured voltage and current would be in the third quadrant of the $R - X$ plane, outside the circle, implying that relay would be blocked from tripping. Such relays are referred as having a "mho" characteristic.

- Pilot Relays: Pilot relays use a communication channel, a power-line carrier or fiber-optics, to communicate between the two terminals of a transmission line being protected by such relays. If a fault occurs that is internal to the transmission line, the relays issue commands to circuit breakers at both ends of this line to interrupt the fault.

13.6.2.2 Zones of Protection in Transmission Lines

The entire power system is divided into zones of protection, for example as shown in Figure 13.22 for the relay at bus-A.

Each zone encompasses one or more power system equipment, and adjacent zones are overlapping with other relays so that no part of the power system is left unprotected even if one or more relays fail to operate.

- Zone 1: The first zone for the relay at A encompasses, for example, 80 percent of the line A-B. If the fault occurs in A's first zone, the relay acts instantly without any time

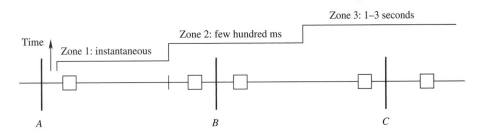

FIGURE 13.22 Zones of protection for the relay at bus-A.

delay. The remaining 20 percent of the line A-B is protected by another relay or relay-element at bus-A, but time-delayed for tripping.

- Zone 2: The second zone for the relay at A encompasses the remaining 20 percent of the line-section A-B, and overreaching in the next section B-C but below 80 percent of the zone-1 reach of the relay at B on line-section B-C. If the fault occurs in the second zone assigned this relay at bus-A, it operates with a time delay of a few hundred msec to prevent its mis-operation. Therefore, this relay must be coordinated with the relay at bus-B, looking towards bus-C, for which 80 percent of the line B-C is its first zone.

- Zone 3: This adjacent zone in this relay at A is set to provide a back-up protection beyond zone 2, for the rest of the line B-C and into the next line, with a time delay of 1 to 3 seconds.

13.6.2.3 Protection of Generators and Transformers

Similar to the bus-protection using a differential relay shown in Figure 13.17, a differential relay shown in Figure 13.23 can provide protection of a generator. A similar scheme can be used for protecting transformers.

Example 13.4

Consider the example three-bus power system shown in Figure 13.24. What types of relays can protect the generator and the transformer at bus-A? What types of relays act on the circuit breaker at bus-A, looking towards bus-B?

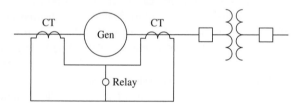

FIGURE 13.23   Protection of a generator using a differential relay.

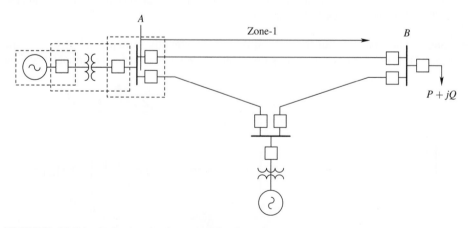

FIGURE 13.24   Relaying in the example three-bus power system.

Solution   The differential protection relay scheme shown in Figure 13.17 and 13.23 can protect the generator, the step-up transformer, and the bus. The circuit breaker at bus-A on line A-B is acted upon by several relays: a differential relay shown in Figure 13.17 to protect the bus, an impedance relay shown in Figure 13.20 to protect against phase-phase faults (not involving ground) between a-b, b-c, and c-a phases and overcurrent relays for single-line to ground (SLG) faults. Only zone 1 of protection is shown in Figure 13.24,

13.6.3   Circuit Breakers

Circuit breakers are large apparatus which, commanded by relays, interrupt the flow of current and break the circuit, as the name implies, in order to protect power equipment from short-circuit faults. There are circuit breakers available that can operate in two cycles and further improvements are continuously being made. Various different principles are used to elongate and cool the arc established by the parting contacts as they try to interrupt the current through them; this process is helped in case of AC circuits where current goes to zero naturally every half-cycle. Circuit breakers use a method of arc interruption based on their voltage level. At 345-kV and above, most circuit breakers use sulfur hexafluoride (SF_6) and SF_6 gas-puffer. SF_6 is a greenhouse gas, but there is no good substitute available.

13.6.3.1   Automatic Reclosure
From stability considerations, high-speed reclosing is important. Since most faults are of transitory nature, many utilities allow EHV circuit breakers a single reclosing automatically. If the fault is still persisting, the circuit breakers trip the line without further attempts to reclose automatically and only the system operator can reclose the line. Such automatic reclosing is not recommended for lines leaving a generating station, since reclosing into a persistent fault can fatigue the shaft of the turbine-generator. In distribution systems, where it is important to maintain continuity of service to customer loads, several attempts for automatic reclosing may be allowed.

13.6.3.2   Single-Phase (Independent-Pole) Operation
In conventional relaying and circuit breaker schemes, all three phases are tripped for any type of fault. However, most faults involve only one of the phases, whereas tripping all three phases represents a more serious interruption. Therefore, certain utilities opt to open only the faulted phase at EHV and UHV levels. Of course, these schemes are more complex and costly than the conventional schemes.

13.6.3.3   Circuit Breaker Ratings
Circuit breakers as discussed above are rated for their interrupting times in cycles of the line frequency. They have voltage ratings based on their insulation and current ratings based on the current they can safely carry and interrupt.

13.6.3.4   Symmetrical and Asymmetrical Current Ratings
A sudden fault in power system can result in current transients that decay based on the X/R ratio of the circuit reactance X to the resistance R. The fault calculations discussed earlier in the chapter determined the *symmetrical* (with equal positive and negative peak values) line-frequency current following a fault. However, with fast circuit breakers it is important to take into account the initial current offset that results in an *asymmetrical*

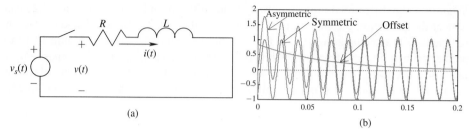

(a)

(b)

FIGURE 13.25   Current in an *RL* circuit.

current, called so because the positive and negative peak values are not the same. The asymmetrical current can be higher than the symmetrical current by a factor depending on the reactance over the resistance ratio, X/R, of the network.

This initial current offset that decays with time can be understood by a simple R-L circuit with a sinusoidal source, as shown in Figure 13.25a, where $v_s(t) = \hat{V}\sin(\omega t + \beta)$.

Choosing the switch-closing time as $t = 0$, the voltage $v(t)$ across the impedance can be expressed as $v(t) = \hat{V}\sin(\omega t + \beta)$. Therefore, in the circuit of Figure 13.25a,

$$Ri + L\frac{di}{dt} = \hat{V}\sin(\omega t + \beta) \quad (t > 0) \tag{13.29}$$

The natural current component, recognizing that this component decays with time, is

$$i_n(t) = Ae^{-t/\tau} \quad (\text{where, } \tau = L/R \text{ is the time constant}) \tag{13.30}$$

The steady-state AC current component in this R-L circuit is

$$i_{AC} = \hat{I}_{AC}\sin(\omega t + \beta - \phi) \tag{13.31}$$

where, $\hat{I}_{AC} = \dfrac{\hat{V}}{\sqrt{R^2 + (\omega L)^2}}$ and $\phi = \tan^{-1}\left(\frac{\omega L}{R}\right)$.

Therefore, the complete current solution, using Equations 13.30 and 13.31, is

$$i(t) = Ae^{-t/\tau} + \hat{I}_{AC}\sin(\omega t + \beta - \phi) \tag{13.32}$$

where the unknown A can be calculated from the initial condition, that is, at time $t = 0$, the current $i = 0$. Therefore, from Equation 13.32, $A = \hat{I}_{AC}\sin(\phi - \beta)$, and thus

$$i(t) = \underbrace{\hat{I}_{AC}\sin(\phi - \beta)e^{-t/\tau}}_{\text{offset } I_{DC}(t)} + \hat{I}_{AC}\sin(\omega t + \beta - \phi) \quad (t > 0) \tag{13.33}$$

This *asymmetrical* current $i(t)$, where the decaying DC offset is superimposed on the AC (symmetrical) component, is plotted in Figure 13.25b. As can be seen from Equation 13.33, the offset $I_{DC}(t)$ is a *DC* component that decays exponentially with time as $e^{-(R/L)t}$, where the circuit time constant $\tau = L/R$. In low-resistance networks that are typical of power systems, the DC component decays slowly compared to the power frequency, and can be assumed constant at the time of the circuit-breaker opening at a value I_{DC} over one or two power-frequency cycles. Therefore, the RMS value I_{rms} of the asymmetrical current can be calculated, recognizing that the RMS value of the AC component is $I_{AC} = \hat{I}_{AC}/\sqrt{2}$, as

$$I_{rms} = \sqrt{I_{DC}^2 + I_{AC}^2} \tag{13.34}$$

The above discussion shows that in low-resistance networks that are typical of power systems, high-speed circuit breakers that interrupt current soon after the fault, have to interrupt a current that is factor S higher than the AC current, where $S = 1.2$ for a two-cycle breaker (this includes ½-cycle minimum relay time and plus the opening time of the breaker contacts) [8]. This factor S varies inversely proportional to the breaker contact parting-time, as defined in the ANSI/IEEE Standard C37.010 Application Guide [8]. For circuit breakers at above 115-kV voltage levels, the closing and latching capabilities for momentary symmetrical rms current are defined to be 1.6 times the rated RMS short-circuit current [9].

REFERENCES

1. C.L. Fortescue, "Method of Symmetrical Coordinates Applied to the Solution of Polyphase Networks," AIEE, vol. 37, pp 1027–1140, 1918.
2. Prabha Kundur, *Power System Stability and Control*, McGraw-Hill, 1994.
3. Paul Anderson, *Analysis of Faulted Power Systems*, IEEE Press, 1995.
4. PowerWorld Computer Program (www.powerworld.com).
5. United States Department of Agriculture, Rural Utilities Service, Design Guide for Rural Substations, RUS BULLETIN 1724E-300 (www.rurdev.usda.gov/RDU_Bulletins_Electric. html).
6. Homer M. Rustebakke (editor) *Electric Utility Systems and Practices*, 4th edition John Wiley & Sons, August 1983.
7. A. Phadke, J. Thorp, *Computer Relaying for Power Systems*, Institute of Physics Publishers, 2005.
8. H. O. Simmons, Jr., "Symmetrical versus Total Current Rating of Power Circuit Breakers," Applications of Power Circuit Breakers, IEEE Tutorial Course, 75CH0975-3-PWR.
9. ANSI/IEEE Standard C37.04.

PROBLEMS

13.1 Due to a single-line to ground fault on phase-a in a system, $\overline{I}_{fa} = 5.0\angle 0$ pu and $\overline{I}_{fb} = \overline{I}_{fc} = 0$. Calculate the symmetrical components $\overline{I}_{fa1}, \overline{I}_{fa2}$ and $\overline{I}_{fa0}$ of the fault current.

13.2 Due to a line-to-line fault between phase-b and phase-c, $\overline{I}_{fa} = 0$ and $\overline{I}_{fb} = -\overline{I}_{fc} = 5.0\angle 0$ pu. Calculate the symmetrical components $\overline{I}_{fa1}, \overline{I}_{fa2}$, and $\overline{I}_{fa0}$.

13.3 At a point f in the system, there is an open-circuit fault on phase-a with a voltage across the open-circuit as $\overline{V}_{fa}$, whereas $\overline{V}_{fb} = \overline{V}_{fc} = 0$. Similar to the calculations of the short-circuit currents by making use of the sequence networks, calculate the sequence components $\overline{V}_{fa1}, \overline{V}_{fa2}$, and $\overline{V}_{fa0}$ at the fault-point.

13.4   Repeat Example 13.2 if the single-line to ground (SLG) fault is through a fault impedance $Z_f = 0.15 \angle 0$ pu.

13.5   Repeat Example 13.2 if before the fault, the load is zero—that is, $P_{Load} = 0$.

13.6   Repeat Example 13.2 if it is a double-line fault between phases b and c.

13.7   Repeat Example 13.2 if it is a double-line fault between phases b and c with a fault impedance $Z_f = 0.15 \angle 0$ pu.

13.8   Repeat Example 13.2 if it is a double-line to ground fault with phases b and c grounded.

13.9   Repeat Example 13.2 if it is a double-line to ground fault with phases b and c grounded through a fault impedance $Z_f = 0.15 \angle 0$ pu.

13.10 Repeat Example 13.2 if there is a single-line to ground fault on phase-a of bus-1. The generator neutral is grounded through a resistance $R_n = 0.10$ pu.

13.11 Repeat Example 13.2 if there is a line-line to ground fault involving phases b and c of bus-1. The generator neutral is grounded through a resistance $R_n = 0.10$ pu.

13.12 Repeat Example 13.2 for an open-circuit fault by calculating the sequence voltages at the fault-point on line 2-3 near bus-2, where the contact opens up on phase-a.

13.13 With an impedance relay with the characteristic shown in Figure 13.21 applied to protect line 2-3 in Example 13.2, calculate the point in the impedance plane for a three-phase fault that is 85 percent of the line-length away from bus-2. Repeat this if the fault is 15 percent of the line-length away from bus-2.

13.14 Prove Equation 13.34.

PROBLEM USING POWERWORLD

13.15 Repeat Example 13.3 for a double-line and a double-line to ground fault.

13.16 In Figure 13.22, draw the protection zones for the relay at bus-B looking toward bus-C.

13.17 In Figure 13.22, draw the protection zones for the relay at bus-C looking toward bus-B.

14

TRANSIENT OVERVOLTAGES, SURGE PROTECTION, AND INSULATION COORDINATION

14.1 INTRODUCTION

Transmission and distribution lines form a large network spread over thousands of miles. A variety of reasons discussed below can cause abnormally high overvoltages that, unless protected properly against, can disrupt service momentarily at best, and cause prolonged outages and expensive damage to the power systems apparatus at worst. In this chapter, we will examine the causes of overvoltages and the measures that can be taken to protect against them.

14.2 CAUSES OF OVERVOLTAGES

Overvoltages are caused primarily by lightning strikes and switching of extra-high-voltage transmission lines, both of which are examined below.

14.2.1 Lightning Strikes

As mentioned earlier, transmission lines, stretched over long distances, are frequently subjected to lightning strikes. Frequency of such strikes depends on their geographical location. Lightning is not a very well understood phenomenon but it would suffice here to say that a lightning strike to an apparatus results in a brief discharge of current pulse to it, with respect to ground. This pulse is commonly represented to reach its peak in time t_1 and exponentially tapers off to half its peak value at t_2. The peak current strokes of as high as 200 kA have been recorded, though generally a peak of 10 kA to 20 kA is assumed to protect against the resulting overvoltages.

14.2.1.1 Lightning Strike to Shield Wires

Many transmission lines have shield-wires as shown in the tower structure in Figure 14.1a. These shield wires are located higher than the transmission line conductors and hence protect them from being hit directly by lightning strokes. They provide an approximately a 30-degree protection zone around them. These shield wires are grounded

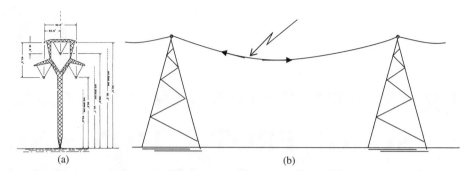

FIGURE 14.1 Lightning strike to the shield wire.

through the tower, where it is desirable to keep the tower footing impedance as small as possible. It should be noted, however, that not all transmission systems have shield wires; many utilities find it more economical to use surge arresters more extensively (although not at every tower) rather than employing shield wires.

When the lightning strikes a shield wire, the resulting current waves flow in both directions and pass through the towers to ground, as shown in Figure 14.1b. The tower footing resistance and the $L(di/dt)$ effect due to the rapidly rising current-front may cause the tower potential to exceed the insulation strength of the insulator string, causing it to flashover (called backflash).

14.2.1.2 Lightning Strike to a Conductor
Another scenario is where one of the conductors is struck by lightning. Resulting traveling waves proceed in both directions and insulator strings may flashover; otherwise, when the current wave reaches the termination, the surge arresters, discussed later, would prevent the voltages from rising to a level that can damage the apparatus.

14.2.2 Switching Surges

At extra-high voltage levels and above, switching of transmission lines, as shown in Figure 14.2a can result in overvoltages, as shown in Figure 14.2b, that can be higher than those caused by lightning strikes. We will see later on that the requirement of the needed insulation is influenced by both the amplitude as well as the duration of the overvoltages. These switching overvoltages can be minimized by the use of pre-insertion resistors in Figure 14.2b, where prior to switching in a transmission line, a resistor in series can be inserted which is bypassed later on for the normal operation.

14.3 TRANSMISSION LINE CHARACTERISTICS AND REPRESENTATION

The discussion of transmission lines here also includes distribution lines. In the following analysis, we will assume a transposed three-phase line and utilize the discussion in Chapter 4 dealing with transmission lines in steady state. Assuming this line to be lossless, the voltage and current magnitudes of these traveling waves are related

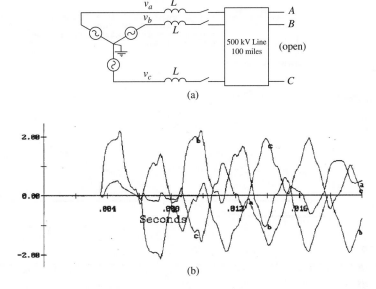

FIGURE 14.2   Overvoltages in per-unit due to switching of transmission lines.

by the characteristic impedance Z_c of this transmission line where, as calculated in Chapter 4,

$$Z_c = \sqrt{\frac{L}{C}} \tag{14.1}$$

and L and C are inductance and capacitance per-unit length of the transposed three-phase lines. These travelling waves propagate at a velocity c that is related to the transmission-line parameters in the following manner:

$$c = \sqrt{\frac{1}{LC}} \tag{14.2}$$

From Chapter 4, the per-phase inductance per-unit length is

$$L = 2 \times 10^{-7} \ln \frac{\sqrt[3]{D_{12}D_{23}D_{31}}}{r} \text{ H/m} \tag{14.3}$$

where the current is assumed to be at the outer surface of the conductor with a radius r. Also from Chapter 4, the per-phase capacitance per-unit length in air is

$$C = \frac{2\pi \times 8.85 \times 10^{-12}}{\ln \dfrac{\sqrt[3]{D_{12}D_{23}D_{31}}}{r}} \text{ F/m} \tag{14.4}$$

Using Equations 14.3 and 14.4, the traveling wave speed in Equation 14.2 is $c = 3 \times 10^8$ m/s, which is the speed of light. This speed is less for practical transmission lines with losses, particularly in the zero-sequence mode.

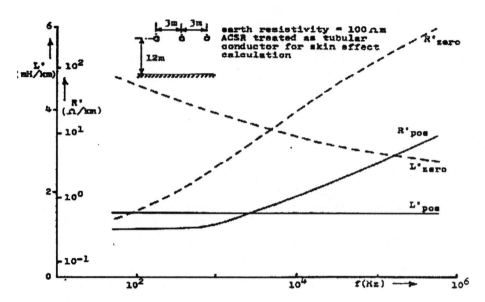

FIGURE 14.3   Frequency dependence of the transmission line parameters [Source: 2].

The transient disturbances are usually not balanced, for example only one of the transmission-line phases is normally subjected to a lightning impulse. Therefore, the zero-sequence path involving ground return must be carefully included in modeling to get correct overvoltages. Similarly, transmission lines are not perfectly balanced and most lines are not transposed, and hence may require to be represented as un-transposed. Moreover, associated with these transient phenomena, we are no longer dealing with a steady state line frequency of 60 or 50 Hz; rather a transient operation that involves very high frequencies. Figure 14.3 [2] illustrates the frequency dependence of the line parameters for positive (and negative) sequence quantities and the zero-sequence quantities as well, and this frequency-dependence should be included for an accurate analysis.

14.3.1   Calculation of Overvoltages

In very simple cases, overvoltages can be calculated by means of keeping track of traveling waves by means of what is called Bewley diagram, after Mr. L.V. Bewley of the General Electric Company who pioneered it, or by means of using Laplace transforms. However, even the simplest three-phase system with un-transposition and frequency dependence of parameters is too complex to be amenable to such methods, where traveling waves in various modes combine to result in the resultant voltage [2]. Calculation of surge voltages due to transmission-line switching are illustrated by means of an example below, using *EMTDC* [3].

Example 14.1

In the three-bus example power system of Chapter 5, the transmission line between buses 1 and 3 is opened at both ends following a fault, as shown in Figure 14.4. It is switched in first at bus-1 end, with the end at bus-3 still open. Calculate the switching overvoltages at the bus-3 end of this transmission line using *EMTDC*.

FIGURE 14.4 Calculation of switching overvoltages on a transmission line.

Solution The modeling of this system and the resulting overvoltages are calculated using EMTDC and the results are included on the accompanying website.

14.4 INSULATION TO WITHSTAND OVERVOLTAGES

Power systems apparatus is designed with insulation to withstand a certain amount of voltages. The insulation level to withstand voltages depends on factors such as the shape of the voltage wave and its duration, condition of the insulation in terms of its age, humidity and the contamination levels. Power systems apparatus can be categorized into having two types of insulations: self-restoring and non-self-restoring. Insulator strings on transmission lines represent the self-restoring type of insulation which can be allowed to breakdown occasionally and restores itself after the resulting fault clears. However, insulation for transformers, generators, and so on is of the non-self-restoring type, which if allowed to fail at any time will result in severe and permanent damage that will be very expensive to repair.

For self-restoring insulation such as the insulator strings of transmission lines, it is cost-prohibitive to insulate against all expected overvoltages and hence a statistical approach may be used, where a certain probability of failure is acceptable. The non-self-restoring insulation of transformers, for example, is protected against failures using surge arresters, as discussed later in this chapter.

As mentioned earlier, the withstand voltage capability of insulation depends on the shape of the voltage wave and its duration, which are different for the lightning and the switching surges. These voltage levels that a given insulation can withstand are defined below.

14.4.1 Basic Insulation Level (BIL)

For the lightning impulse, the level of insulation is specified as the basic insulation level (BIL), which is the peak of the withstand voltage of a standard lightning impulse voltage wave. This lightning impulse wave, as shown in Figure 14.5, is assumed to rise to its peak value in $1.2\,\mu s$, and tapers off exponentially to one-half of its peak value in $50\,\mu s$.

The basic insulation level (BIL), as an example, for a 345-kV transformer is shown in Figure 14.6 as 1175 kV, which is the peak line-to-ground voltage.

14.4.2 Basic Switching Insulation Level (BSL)

As shown in Figure 14.2, voltages applied at one end of a transmission line result in traveling waves of voltage and current with steep fronts. These switching surges do not have as steep a front as those resulting from lightning impulse; however, these switching surges are of longer duration. The standard switching impulse voltage wave is assumed to

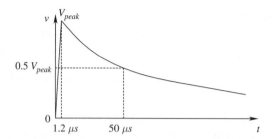

FIGURE 14.5   Standard Voltage Impulse Wave to define BIL.

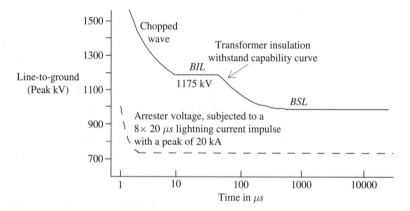

FIGURE 14.6   A 345-kV transformer voltage insulation levels.

rise to a peak value in 250 μs and tapers off exponentially to one-half of its peak value in 2500 μs. The withstand insulation level for the peak of such as a switching surge is called the basic switching insulation level (*BSL*). As shown in Figure 14.6, an apparatus with a particular BIL has a lower withstand capability to switching surges—that is, it has a lower BSL than BIL due to the fact that the switching surges are of a longer duration.

14.4.3   Chopped-Wave Insulation Level

As shown in Figure 14.6, the same apparatus insulation can withstand higher voltages than its BIL if the voltage impulse applied to this apparatus is chopped to bring it to zero in 3–4 μs after the impulse has reached its peak value.

14.5   SURGE ARRESTERS AND INSULATION COORDINATION

In power systems apparatus, their insulation is protected against voltage impulses by means of surge arresters connected in parallel to them. The role of surge arresters is somewhat similar to that of Zener diodes in low-power electronic circuits. Surge arresters appear as an open-circuit at normal voltages and draw almost negligible current. However, by providing a low-impedance path, they allow whatever current needs to flow, of course within their rating, by proper selection, thus "clamping" the voltage across the device being protected to a threshold voltage (plus a small *IR* current-discharge voltage), as shown in Figure 14.6 by the dotted curve. A surge arrester should be able to dissipate the associated energy following an impulse, without any damage to itself, thus allowing a quick return to the normal operation when it should once again appear as an open-circuit. In power systems, modern practice is to use metal oxide (zinc oxide) arresters which have a highly nonlinear *i-v* characteristic, for example, as given by the following relationship:

$$i \propto V^q \tag{14.5}$$

where the voltage is expressed in per unit of the overprotection voltage value and the exponent q can be as high as 26 to 30. Therefore, the current through the arrester for voltages substantially below unity is negligible. However, in the overvoltage protection region, the arrester voltage remains almost constant. The arrester discharge voltage level should be sufficiently below the apparatus insulation level that this arrester is to protect, and provide a margin given by various IEEE and ANSI standards. Therefore, the insulation coordination requires coordinating and selecting BIL and BSL of various apparatus and the ratings of the arresters to protect them, as illustrated in Figure 14.6.

REFERENCES

1. Electric Power Research Institute (EPRI), *Transmission Line Reference Book: 345* kV *and above*, 2nd edition.

2. Hermann W. Dommel, *EMTP Theory Book*, BPA, Contract No. DE-AC79-81BP31364, August 1986.

3. PSCAD/EMTDC, Manitoba HVDC Research Centre: www.hvdc.ca.

4. Surge Protection in Power Systems, IEEE Tutorial Course, 79EH0144-6-PWR.

5. United States Department of Agriculture, Rural Utilities Service, Design Guide for Rural Substations, RUS BULLETIN 1724E-300 (www.rurdev.usda.gov/RDU_Bulletins_Electric.html).

PROBLEMS

EMTDC-Based Problem

14.1   As described on the accompanying website, calculate the positive-sequence and the zero-sequence impedances of the transmission line in Example 14.1.

14.2   As described on the accompanying website, calculate the switching surges with and without the use of pre-insertion resistors.

14.3   As described on the accompanying website, calculate the switching surges with and without the use of surge arresters at the open-end.

INDEX

237